TRANSPORT AND SURFACE PHENOMENA

From the Czech original translated by **J. J. ULBRECHT**

TRANSPORT AND SURFACE PHENOMENA

KAMIL WICHTERLE
MAREK VEČEŘ

ELSEVIER

Elsevier
Radarweg 29, PO Box 211, 1000 AE Amsterdam, Netherlands
The Boulevard, Langford Lane, Kidlington, Oxford OX5 1GB, United Kingdom
50 Hampshire Street, 5th Floor, Cambridge, MA 02139, United States

Notices

Knowledge and best practice in this field are constantly changing. As new research and experience broaden our understanding, changes in research methods, professional practices, or medical treatment may become necessary.

Practitioners and researchers must always rely on their own experience and knowledge in evaluating and using any information, methods, compounds, or experiments described herein. In using such information or methods they should be mindful of their own safety and the safety of others, including parties for whom they have a professional responsibility.

To the fullest extent of the law, neither the Publisher nor the authors, contributors, or editors, assume any liability for any injury and/or damage to persons or property as a matter of products liability, negligence or otherwise, or from any use or operation of any methods, products, instructions, or ideas contained in the material herein.

Library of Congress Cataloging-in-Publication Data
A catalog record for this book is available from the Library of Congress

British Library Cataloguing-in-Publication Data
A catalogue record for this book is available from the British Library

ISBN: 978-0-12-818994-8

Publisher: Susan Dennis
Acquisitions Editor: Anneka Hess
Editorial Project Manager: Sara Valentino
Production Project Manager: R.Vijay Bharath
Cover Designer: Mark Rogers

Typeset by SPi Global, India

Contents

Foreword

The three basic engineering tasks

The fundamental task of an engineering assignment is to find qualitatively possible solutions to a given problem and to answer the quantitative question about its technical feasibility and economic advantage.

The qualitative solution usually leans on the search of the scientific, commercial, and patent literature but, in the first place, on one's own experience combined with one's own capability of combinations and fantasy.

When a quantitative solution is required then the engineer is asked to apply the three basic engineering tasks, often using simple, but more often sophisticated mathematical methods.

The first task is based on simple, and in the simplified world of real life, undisputed principles of **natural sciences**, such as the law of conservation of mass, the law of conservation of various kinds of energy, and the stoichiometric laws of chemistry; also included are simple rules for the conversion of quantities such as the mass and volumes. Using these rules one can reliably perform **mass and energy balances** of a particular process.

The second task follows a question how far can a certain process proceed, or under what circumstances will the process change qualitatively. In this case, one can make use of known **process equilibrium**. The quantities that enter into the calculations here are more complex parameters from the domain of **statics** and **thermodynamics** that can have the form of constants, functions, or complex mathematical objects, the values of which must be determined experimentally. Alternatively, the values of some (such as the melting points, vapor tensions, solubilities, interfacial tensions, and bonding enthalpies) can be found with sufficient accuracy in a tabular form in the appropriate literature. In some cases, one can use relatively reliable formulas for the estimation of these quantities into which one needs to substitute quantities that are easier to obtain.

The third task deals with the **kinetics** of a process, i.e., what is its rate. In the case of chemical and physical processes that proceed outside the equilibrium, the rate is determined by the set of driving forces opposed by a set of resistances that act against the equalization of the properties in different parts of the system. The three most significant resistances are those that oppose the **transfer of momentum**, **the transfer of heat**, and **the transfer of mass**. These resistances have, to a certain degree, a linear character so it is advantageous to study them together under a common title of **transport phenomena**. A separate task is the study of the **chemical kinetics** that, as a rule, is not linear and, therefore, must be studied independently.

Transport phenomena: The continuum and the interfaces

Under ordinary circumstances we perceive through our senses any mass as being a **continuum** or as a set of continua separated by simple **geometrical interfaces**. In reality (unless one wants to immerse deeper into nuclear physics), the mass is rather a set of moving particles (vibrating, sometimes rotating, and even mutually exchanging their positions). Despite of that, it is useful to consider a large set of particles (be they molecules, nanoparticles, or even particles visible by bare eye) to be a continuum to which one can assign **averaged quantities** to every point of the system. **Intensive quantities** are ratios of directly measurable extensive quantities (e.g., the density is the ratio of mass and volume at a given point).

Among the **statistically averaged quantities**, the local velocity is the statistical average of particle paths cutting the imaginary plane in the vicinity of the point in question per unit of time) or the temperature (expressing mean energy of the oscillating molecules).

To describe such an approach, the **classical physics** developed mathematical tools; the analysis of continuous functions (Newton, Leibnitz) lead to usual **differential and integral calculus**.

The equilibrium states in homogeneous phases are characterized by invariability of quantities that can be called the **measures of equilibrium**, for example, the temperature. When the temperature is constant, there is no local heat transport, when the concentration (activities, fugacities) are constant, there is no mass transport, etc. The discipline of physics that deals with the usual equilibrium processes is called statics; in physical chemistry it is called thermodynamics.

When the system is **not in equilibrium** then the variability of those quantities that measure the equilibrium lead to position changing, regrouping, and transformations of both mass and energy. To evaluate the rate with which the system moves toward the equilibrium, the so-called **transport processes**, it is useful to introduce quantities that characterize the basic transport phenomena: **momentum transfer, heat transfer, and mass transfer**. The relevant transfer coefficients are **viscosity, thermal conductivity, and diffusivity**.

In the past, the individual phenomena were studied in separate scientific disciplines, such as the fluid mechanics, hydrodynamics, and the science of heat and mass transfer. Later, it turned out, however that these three processes are analogous and that they can be treated with the same mathematical tools.

As a rule, the individual steps of an engineering solution usually are as follows:

- The definition of a spatial and temporal region in which the process in question takes place.
- The application of a mathematical model to the transport in question including boundary and initial conditions.
- An analysis of possible solutions of the model, first in a simplified form.
- The evaluation of all possible solutions, in simple cases, is carried out.
- The presentation of the results in a useful form.

The transport across a phase boundary is a special case. The difference between the transport coefficients in different mediums results in a discontinuity that modifies velocity fields, velocity gradients, temperature gradients, concentration gradients, etc. The opposing limits at the phase boundary are usually related by a simple mathematical formula that is fortunately not modified by the transport itself. Therefore, the common rules of physics and physical chemistry describing the equilibriums at phase boundary remain valid. Thus, the knowledge of surface phenomena is essential for the study of complex transport phenomena.

The history of this discipline

A perfunctory opening of any monograph titled "Transport Phenomena" (possibly expanded in "Transport of momentum, heat, and mass") might lead the reader to erroneous conclusions. Those who are not too fond of mathematics are frightened by the mass of formulas and relationships from the domain of differential equations and those from the vector count. On the other hand, mathematical geeks see a fascinating playing field.

This should not be very surprising since this discipline grew out of the work of such **physicists** (Newton, Bernoulli, Euler, Cauchy, Stokes, and others) who are known by their discoveries in mathematics needed for solutions to specific physical problems. It is even possible to claim with a certain degree of exaggeration that **mathematics** in its time owed its progress more to the study of transport phenomena than the science of transport phenomena to mathematics. Up to the end of the 19th century, these disciplines formed relatively special and not too emphasized part of physics.

Next came the period characterized by the progress in ship transportation, upswing of aviation, steam energy generation, and chemical engineering, when this discipline becomes the subject of interest of experimentalists and mathematically trained engineers such as Nusselt, Kirchhoff, Reynolds, Grashof, Zhoukovsky, Prandtl, Karman, Luikov, Kolmogoroff, and others.

In 1956, Professor Hans Kramers from the Technical University, Delft (The Netherlands) for the first time connected and organized the existing knowledge about the transports of momentum, heat, and mass in a series of lectures "Fysische Transportverschijnselen" so that methods hitherto used only in particular cases could now be used generally. This idea was further expanded by his disciple R.B. Bird, later professor of chemical engineering at the University of Wisconsin, USA, and his colleagues Professors W.E. Stewart and E.N. Lightfoot in the seminal monograph "Transport Phenomena" published in 1960.

In the next two decades, the discipline became a domain of **theoreticians** from universities who gradually filled any remaining gaps in already solved tasks. In most cases, these were theoretical computational works even though some of them were just insignificant problems. Still, the appearance of new advanced experimental methods made it possible to carry out a series of new measurement techniques. The inclusion of transport phenomena courses in engineering curricula at universities became common.

With the exponential growth of the performance of the computational techniques, with the expansion of personal computers, and with the appearance of user-friendly software has diminished the pressure on the **mathematical** skills of the submitter. Transport phenomena become again available to **engineers** and generally for chemical engineers who need to understand the processes taking place under more complex conditions since they have to design these processes. The tempting availability of ready-made solutions should not, however, distract from a sensible definition of the problem and from the rational specification of the main goal of the calculations.

The structure of this book

This book originated from the lectures given at the Technical University Ostrava (Czech Republic) in 1995 for graduate students of chemical engineering. The lectures were also attended by graduate students of material engineering, environmental engineering, safety engineering, and nanotechnology. Therefore, the book is divided into three parts:

In the first part named "Fundamentals of Transport Phenomena" the focus is on geometrically simple setups in which the transport occurs. Some more common problems that are often encountered by physical chemists, material engineers and technologists (metallurgists, plastics engineers, ceramicists, food processing engineers, etc.) are expanded. To a large extent, this part is only qualitative but it defines certain fundamental concepts of this discipline that are later used in more complex situations.

The second part named "Balances of Transport Phenomena" is intended for practicing hydrodynamicists, heat technologists, and generally to chemical engineers who need to understand the processes under more complex conditions and design such processes. The solution to these tasks usually requires more complex mathematical apparatus. Therefore, it is the basic priority of this work to get the student formulate a specific task to such an extent that the actual solution can be performed with the help of the computational methods by a specialist who does not need to know much about the physical nature of the problem. A complete solution usually leads to the description of the velocity, temperature, or concentration fields but the engineer needs to obtain out of these only relatively simple results, for example, only some local values or those suitably averaged. The final aim is thus is to suitably specify the task and suitably interpret the results.

The third part termed "Mathematical Methods for Solving the Transports" concisely elucidates the mathematical means used to solve the transports: vector and tensor count and the selected approaches to solving partial and ordinary differential equations that do not belong to the common widely trained gear of college students of technical and science faculties.

The aim of this text was not the creation of an encyclopedic treatment of the subject neither to stay with a simple manual for the solution of a few standard tasks. The aim is to demonstrate how the necessary concepts and methods of the discipline were created and to show what are they useful for. The student should become competent in understanding original works dealing with the transfer of momentum, heat, and mass and orient him/herself in numerous monographs that systematically deal with the subject in various depths and extent and with miscellaneous orientation and application.

For the price of a certain inconsistency it will be attempted not to overwhelm the student with an excess of formalisms. Right from the beginning the text is sandwiched between notes referring to practical problems in which the discussed theoretical concepts are applied.

Other books

Van den Akker, H., Mudde, R.F., 2014. Transport Phenomena—The Art of Balancing, first ed. Delft Academic Press/VSSD, Delft, The Netherlands.

Asano, K., 2006. Mass Transfer: From Fundamentals to Modern Industrial Applications. Weinheim, Wiley VCH, Germany.

Astarita, G., Ocone, R., 2002. Special Topics in Transport Phenomena. Elsevier, Amsterdam, New York, United States.

Baukal, C.E., 2000. Heat Transfer in Industrial Combustion. CRC Press, Boca Raton, United States.

Beek, W.J., Muttzall, K.M.K., Van Heuven, J.W., 1999. Transport Phenomena, second ed. John Wiley & Sons, Inc., New York, United States.

Belfiore, L.A., 2003. Transport Phenomena for Chemical Reactor Design. J. Wiley, New York, United States.

Bennett, C.O., Meyers, J.O., 1982. Momentum, Heat and Mass Transfer, McGraw-Hill College subsequent ed. New York, United States.

Bergman, T.L., 2011. Introduction to Heat Transfer. John Wiley & Sons, Inc., Hoboken, NJ, United States.

Bird, R.B., Stewart, W.E., Lightfoot, E.N., 2006. Transport Phenomena, revised second ed. John Wiley & Sons Inc., New York, United States.

Bird, R.B., Stewart, W.E., Lightfoot, E.N., Klingenberg, D., 2015. Introductory Transport Phenomena. John Wiley & Sons Inc., New York, United States.

Brodkey, R.S., Hershey, H.C., 2003. Transport Phenomena: A Unified Approach. Brodkey Publishing, Columbus, United States.

Carslaw, H.S., Jaeger, J.C., 1959. Conduction of Heat in Solids. Clarendon Press, Oxford, UK.

Çengel, Y.A., Ghajar, A.J., 2015. Heat and Mass Transfer: Fundamentals & Applications. McGraw-Hill, New York, United States.

Cussler, E.L., 2009. Diffusion: Mass Transfer in Fluid Systems. Cambridge University Press, Cambridge, New York, United States.

Das, M.K., Mukherjee, P.P., Muralidhar, K., 2018. Modeling Transport Phenomena in Porous Media With Applications. Springer International Publishing.

Deen, W.M., 1998. Analysis of Transport Phenomena. Oxford University Press, Oxford, England, UK.

Faghri, A., Zhang, Y., 2006. Transport Phenomena in Multiphase Systems. Elsevier, New York, United States.

Farmer, R.C., Pike, R.W., Cheng, G.C., Chen, Y.-S., 2009. Computational Transport Phenomena for Engineering Analyses, first ed. CRC Press, New York, United States.

Fournier, R.L., 2017. Basic Transport Phenomena in Biomedical Engineering, fourth ed. CRC Press, Boca Raton, United States.

Geankoplis, CH.J., 1993. Transport Processes and Unit Operations, third ed. PTR Prentice-Hall, Inc., A Simon & Schuster Company, Englewood Cliffs, New Jersey, United States.

Geankoplis, CH.J., Hersel, A.A., Lepek, D.H., 2018. Transport Processes and Separation Process Principles, fifth ed. Prentice Hall International Series, New Jersey, United States.

Geiger, G.H., Poirier, D.R., 1973. Transport Phenomena in Metallurgy, first ed. Addison-Wesley, Reading, United States.

Greenkorn, R., 1999. Momentum, Heat, and Mass Transfer Fundamentals, first ed. CRC Press, New York, United States.

Griskey, R., 2006. Transport Phenomena and Unit Operations. A Combined Approach. John Wiley & Sons Inc., New York, United States.

Hanjalic, K., Kenjeres, S., Tummers, M.J., Jonker, H.J.J., 2009. Analysis and Modelling of Physical Transport Phenomena. VSSD, Delft, The Netherlands.
Hauke, G., 2008. An Introduction to Fluid Mechanics and Transport Phenomena. Springer, Netherlands.
Incropera, F.P., 2007. Fundamentals of Heat and Mass Transfer/Frank P. Incropera ... [et al.], John Wiley, Hoboken, NJ, United States.
Jakobsen, H.A., 2014. Chemical Reactor Modeling: Multiphase Reactive flows. Springer, Berlin, Germany.
Jiji, L.M., 2009. Heat Convection. Springer-Verlag, Berlin, Heidelberg, Germany.
Jiji, L.M., 2009. Heat Conduction. Springer-Verlag, Berlin, London, Germany.
Kou, S., 1996. Transport Phenomena in Materials Processing. John Wiley & Sons Inc., New York, United States.
Leal, G.L., 2007. Advanced Transport Phenomena, Fluid Mechanics and Convective Transport Processes. Cambridge University Press, Cambridge, England, UK.
Levenspiel, O., 2014. Engineering Flow and Heat Exchange. Springer, New York.
Lue, L., 2014. Momentum, Heat, and Mass Transfer. Bookboon (Bookboon.COM).
Mashelkar, R.A., 1989. Transport Phenomena in Polymeric Systems, vol. 2. Balaji International, India.
McComb, W.D., 1990. The Physics of Fluid Turbulence. Oxford University Press, New York, United States.
Plawsky, J., 2014. Transport Phenomena Fundamentals, third ed. CRC Press, New York, United States.
Poirier, D.R., Geiger, G.H., 2016. Transport Phenomena in Materials Processing. Springer International Publishers, Switzerland.
Suryavanshi, B.M., Dongre, L.R., 2006. Transport Phenomena for Chemical, Petrochemical and Polymer Engineering. Nirali Prakashan, Puna, India.
Ramachandran, P.A., 2014. Advanced Transport Phenomena: Analysis, Modeling, and Computations, first ed. Cambridge University Press, Cambridge, England, UK.
Rebay, M., Kakaç, S., Cotta, R.M., 2016. Microscale and Nanoscale Heat Transfer: Analysis, Design and Application. CRC Press, Boca Raton, FL, United States.
Rohsenow, W.M., Choi, H.Y., 1961. Heat, Mass, and Momentum Transfer. Prentice-Hall, Englewood Cliffs, New Jersey, United States.
Rorrer, G.L., Foster, D.G., Welty, J., 2014. Fundamentals of Momentum, Heat, and Mass Transfer, revised sixth ed. John Wiley & Sons Inc., New York, United States.
Ruocco, G., 2018. Introduction to Transport Phenomena Modeling. Springer International Publishing.
Saatdjian, E., 2000. Transport Phenomena: Equations and Numerical Solutions. John Wiley & Sons Inc., New York, United States.
Sáez, A.E., Baygents, J.C., 2014. Environmental Transport Phenomena, first ed. CRC Press, Boca Raton, United States.
Sharma, K.R., 2010. Transport Phenomena in Biomedical Engineering: Artificial Organ Design and Development and Tissue Engineering. McGraw-Hill, New York, United States.
Slattery, J.C., 1972. Momentum, Energy and Mass Transfer in Continua. McGraw Hill Book Co., New York, United States.
Soto, R., 2016. Kinetic Theory and Transport Phenomena. Oxford University Press, Oxford, England, UK.
Thomas, W.J., 2000. Introduction to Transport Phenomena. Prentice Hall, Upper Saddle River, New Jersey, United States.
Tosun, I., 2007. Modeling in Transport Phenomena: A Conceptual Approach, second ed. Elsevier, Amsterdam, The Netherlands.
Truskey, G., Yuan, F., Katz, D., 2007. Transport Phenomena in Biological Systems: International edition. Pearson, London, England, UK.

Venerus, D.C., Öttinger, H.Ch., 2018. A Modern Course in Transport Phenomena. Cambridge University Press, Cambridge, England, UK.
Welty, J.R., Wicks, Ch.E., Wilson, R.E., 1976. Fundamentals of Momentum, Heat, and Mass Transfer, second ed. John Wiley & Son, New York, United States.
Welty, J.R., Rorrer, G.L., Foster, D.G., 2014. Fundamentals of Momentum, Heat, and Mass Transfer, revised sixth ed. John Wiley & Sons, New York, United States.
Yang, W.-J., Mochizuki, S., Nishiwaki, N., 1994. Transport Phenomena in Manufacturing and Materials Processing, a Volume in Transport Processes in Engineering. Elsevier Science. Amsterdam, The Netherlands.

Acknowledgments

We wish to thank J.J. Ulbrecht who, 55 years ago lent his student, Kamil Wichterle, his copy of Bird-Stewart-Lightfoot's *Transport Phenomena*. At the age of 90, professor Ulbrecht offered to translate our lecture notes from Czech to English and he succeeded to complete the project within 1 year.

PART ONE

The nature of transport phenomena

CHAPTER ONE

Definitions of unidirectional steady transport

Contents

The simplest cases of transports are demonstrated in thought experiments when there is only one imbalance manifested itself by changes in only one property in one direction. Then a coordinate system can be set up in such a way that the equilibrium measures have a constant value in x_0 and x_1 planes, which is shown schematically in Fig. 1.1A. Then a Cartesian coordinate system x, y, z can be set up to carry out the transport only in the z direction between x_0 and x_1 through the area $\Delta S = \Delta y \Delta z$. For the time being, let us ignore the way this status has been reached and maintained and what is going on outside the system just defined.

Daily experience shows that, at least for some transports, it is possible to find a **linear relationship** between the **intensity of the transfer** (i.e., the transfer through a unit area over a unit of time) and the **driving force** (i.e., the difference between the measures of a particular balance). This can be first demonstrated for the transfer of heat energy.

1.1 Steady unidirectional transport of heat by conduction

In the case shown in Fig. 1.1A, there will be two constant, timely invariable temperatures T_0 and T_1 in the planes $x < x_0$ and $x > x_1$. The history of how this state was arrived at as well as what is going on at $x > x_0$ and

Transport and Surface Phenomena
https://doi.org/10.1016/B978-0-12-818994-8.00001-4

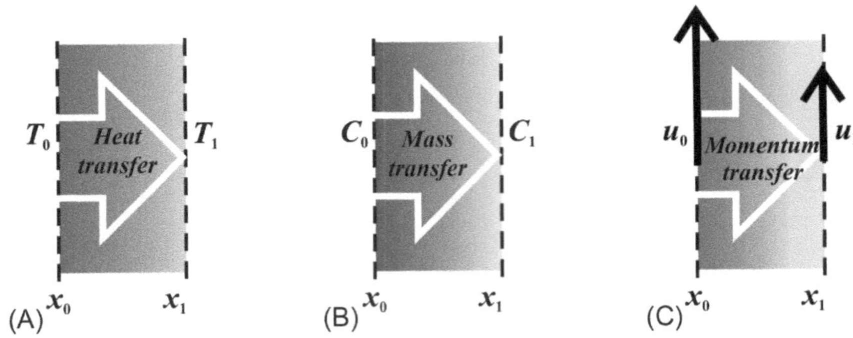

Fig. 1.1 Transport phenomena: (A) heat, (B) mass, (C) momentum.

$x < x_1$ will be ignored. Should the temperatures T_0 and T_1 be are both different then energy transport in the form of **heat** Q (the unit being Joule J) will occur at a rate (heat transport) dQ/dt (the unit being J/s, i.e., Watt, W). It is obvious that this transport will be directly proportional to the size of the area ΔS. It is therefore useful to introduce a quantity q by the definition

$$q \equiv \frac{1}{\Delta S} \frac{dQ}{dt} \tag{1.1}$$

where q is the heat transported through a unit area during a unit time [J/(m^{2}s), i.e., W/m^2] called the **intensity of the heat flow**, alternatively the **density of the heat flow**. The intensity of the heat flow depends on the **driving force** $(T_0 - T_1)$. There are several mechanisms by which the heat can be transported. In the case of small temperature differences in a stationary matter, the key mechanism is the **conduction of heat**, where q is directly proportional to the driving force and indirectly proportional to the **thickness** of the material $h \equiv (x_1 - x_0)$, thus

$$q = k \frac{T_0 - T_1}{h} \tag{1.2}$$

where the constant k defined by this relation is one of the **material constants** that depends only on the material properties, i.e., on the **composition** and the **structure** of the material as well as on the **physicochemical state of aggregation** of the matter. This material constant that is called **thermal conductivity** has the dimension of W/(mK). The relation (1.2) can be expressed also in a differential form as

$$q_x = -k \frac{dT}{dx} \tag{1.3}$$

known as the **Fourier law** of unidirectional thermal conductivity. The minus sign indicates that the heat moves from warmer plane to cooler one.

1.2 Steady unidirectional transfer of mass by diffusion

An analogous situation is shown in Fig. 1.1B where the control areas are kept at different **concentrations**. Also, for this case, a linear relationship can be formulated between the **intensity** (density) of the transfer (i.e., the flow through a unit of area over a unit of time) and a **driving force** (i.e., the difference between the measures of a particular balance).

Several different units can be used to express the transfer of mass (material or volume). In the simplest case, material units can be used to express the transfer of material A. If the **material amount** is denoted as N (the basic unit is 1 mol) then the material flow (rate) is dN/dt (the unit being mol/s). Then the **intensity of the material flow** [measured in mol/(m^{2}s)] is expressed as

$$n \equiv \frac{1}{\Delta S} \frac{dN}{dx} \tag{1.4}$$

which is directly proportional to the driving force that, in this case, will be the **concentration difference** ($C_0 - C_1$) (measured in mol/m^3). It is obvious that the flow of the material will be indirectly proportional to the **thickness** of the layer h. The constant of proportionality $\mathcal{D}$ known as the **diffusivity** is a material constant, the dimension of which is m^2/s.

It would be more correct to use activities rather than concentrations but, in most practical situations, activities are almost equal to concentrations. Partial pressures (or rather fugacities) are more suitable for mass transfer in gases then, however, state equations ought to be employed to calculate the driving force. Similarly, if it is more useful to express the intensity of flow in other quantities than are the mass amounts, then the relevant correlations are to be used to obtain the mass flow or the transfer of volumes. For these recalculations, the fundamental equations in terms of mass concentrations and flows have to be used.

The **diffusion coefficient** is a more complex material constant since it depends on the properties of both components, that is, the **diffunding molecule** as well as the **medium** in which the molecule moves. By analogy with the conduction of heat, the defining relationship for the unidirectional diffusion can be expressed as

$$n_x = -\mathcal{D}\frac{dC}{dx}_x \qquad (1.5)$$

which is known as **Fick's law**.

1.3 Steady unidirectional transfer of momentum by viscous friction in fluids

Somewhat more complex is the situation depicted in Fig. 1.1C in a layer of fluid. In this case, there are different tangential velocities on the two boundaries u_y in the y direction. This configuration is called the simple tangential flow and the transfer of momentum is mediated by a tangential force F_{xy} that has a tendency to speed up the slower layer while slowing up the faster one. According to the second law of motion, the change of momentum will be

$$\frac{d(m\,u_y)}{dt} = F_{xy} \qquad (1.6)$$

It is a matter of convention, which of the two actions will be taken as the positive one. In the classical mechanics, the traction (the pull) is chosen as the positive force so that the **intensity of the momentum transfer** is equal to

$$-\tau_{yx} = \frac{F_{xy}}{\Delta S} \qquad (1.7)$$

Thus, the transfer of momentum is the negative value of the force acting on a unit of area, the so-called **tangential tension** τ_{yx}, the unit of which is the same as the unit of pressure, which is N/m^2 or Pa. It was Newton who for the first time expressed a conjecture that there is a proportionality

$$\tau_{yx} = \mu\,\frac{du_y}{dx} \qquad (1.8)$$

This relationship known as **Newton's law of viscosity** defines a material constant μ, the unit of which is Pa s or kg/m s, which is called the **dynamic viscosity**. The term (du_y/dx) is called **shear rate** or the **rate of deformation**. The often used term "velocity gradient" should be avoided because the shear velocity is same as the velocity gradient only in the simplest constellations (as explained in Chapter 8).

1.4 Similarities and differences

1.4.1 Common features of some transport phenomena

It was shown that the mathematical models of the unidirectional transports of heat, mass, and momentum are similar. The reason for this similarity is that the motion of molecules plays a key role in these transports as shown in the Chapter 2. Even in rather complex cases in which more complicated mathematical tools have to be used, some similarities do exist. Solutions developed for one of these transports can be usually used for the other ones.

1.4.2 Other transport phenomena

Other transport phenomena, taking place over a period of time, are encountered both in the nature as well as in engineering practice. Let us mention in the first place the **transfer of pressure**. This transfer takes place at the speed of sound so it is, with the exception of acoustic and shock waves, rarely perceivable. Even faster is the **transfer of energy by radiation** that takes place at the speed of light. Radiation transport does not need the presence of molecules so it takes place even in vacuum. Later in this book, the **radiation of heat** is briefly treated because it might become comparable with heat conduction.

There are a large number of phenomena due to the **transfer of an electric charge**. This transfer is realized either by the motion of electrons that have negligible inertial mass so they can be put to motion in electric conductors at a infinitesimally short time or by significantly slower motion of ions in electrolytes that is comparable to diffusion. Their motion (electric current due to the difference in electric potentials) is limited by the impedance of the medium that is made of electric resistance and, in particular at the transition between conductors and electrolytes, by inductance and capacity.

1.5 Summary

This chapter deals with three transfer phenomena: momentum, heat, and mass. These transfers are characterized by transfer coefficients that determine the relationship between the flow and the driving force. These transfer coefficients are dynamic viscosity, thermal conductivity, and diffusivity, respectively. The transfer coefficients are material constants in that they depend only on the nature of the media, which is viewed as a continuum.

In unidirectional transfers it holds that

$$\text{Flow} = \text{transfer coefficient} \times \text{driving force}.$$

Table 1.1 summarizes the individual cases.

Table 1.1 Transport and driving force.

Phenomena	Flow	Transfer coefficient	Driving force
Momentum transfer	Tangential stress *Tensor* (Tangential force on unit area) $\boldsymbol{\tau}$ [Pa]	Dynamic viscosity *Scalar* μ [Pa s]	Shear rate *Tensor* $\boldsymbol{\gamma} \equiv (\nabla \boldsymbol{u} + \boldsymbol{u}\nabla)$ [s^{-1}] (Equal to the velocity gradient $\boldsymbol{\gamma} \equiv \nabla \boldsymbol{u}$ [s^{-1}] only in the simplest case !)
Heat transfer	Intensity of heat flow *Vector* q [W]	Thermal conductivity *Scalar* k [W/m K]	Temperature gradient *Vector* ∇T [K/m]
Mass transfer	Intensity of mass flow *Vector* $\boldsymbol{n}$ [mol/(m^2 s)]	Diffusivity *Scalar* $\mathcal{D}$ [m^2/s]	Concentration gradient *vector* ∇C [mol/m^4]

CHAPTER TWO

Transport phenomena in terms of mass structure

Contents

Transport and Surface Phenomena
https://doi.org/10.1016/B978-0-12-818994-8.00002-6

The hypothesis that mass is not continuous, intuitively felt for a long time, was confirmed around the year 1800. It was chemistry that proved that the smallest molecules were of the order of 0.1 nm, that is 10^{-10} m. The dualistic approach of seeing the mass, both as a particulate matter when working with reacting molecules as well as a continuum, when considering volumes or concentrations, is quite common in chemistry to consternation of philosophers. In essence, the approach chosen depends on the scale of consideration. The transition between the particulate approach and the continuous one is in the region of 10–100 nm. The hypothesis about the continuity of mass in an object can be applied without any detriment to the accuracy if the number of particles in the object is very large.

Finding the relationship between the structure of mass and the transport coefficients is also the subject of physical chemistry. The values of viscosity, thermal conductivity, and diffusivity in gases composed of simple molecules can be determined using the molecular kinetic theory of gases. Here, the determining quantities are statistical variables such as the mean free path of molecules and their speed that depend on temperature, pressure, and the molar mass in a known way. Further details can be learned from standard textbooks of physical chemistry.

This matter is somewhat more complex in the case of liquids and solids. However, even in the case of substantially larger macromolecular materials and colloids in which the particle size is of the order of tens or even hundreds nanometers, dependence can be found on the structure and on the state variables. The interparticle attractivity may even include the electrostatic forces. Alternatively, even larger volumes of micrometric or coarse suspensions and emulsions can be characterized as multiphase or quasi-continuous systems. In these systems, viscosity, thermal conductivity, and diffusivity can be at least estimated while studying the transfer of momentum, heat, and mass. Inversely, conclusions can be made about the structure and the constellation of particles in a mass sample from experimentally obtained transfer coefficients.

Thus physical chemistry provides the knowledge of

- how to evaluate the transfer coefficients from simple measurements,
- how the transfer coefficients depend on the state variables,

- how the transfer coefficients depend on the structure of mass,
- what can be learned from the transport processes, and
- how to modify the transfer coefficients in materials.

2.1 Molecular theory of transport phenomena in gases

Let us outline here some of the key theses of the molecular theory of gases without going into their derivation and let us see what they have to offer to the study of transport phenomena.

More details can be found, e.g., in Moore's textbook (Moore, 1999).

2.1.1 Mean velocity of random motion and the temperature

According to the molecular kinetic theory, the gas molecules not only display a statistically averaged motion that, in a continuum, presents itself as a **flow** but also an outwardly unobservable **motion** that is dimensionally **random**. In a continuum, this motion is manifested as the **temperature**. The energy of this motion for small molecules is derived mainly from the speed of translations in all directions characterized by the average mean absolute **velocity of the random motion** c. Its relationship to the temperature can be expressed by

$$c \sim \sqrt{\frac{R\,T}{M}} \tag{2.1}$$

Symbol "∼" denotes a direct proportionality that could be replaced by an operator "=K×," where the dimensionless coefficient of proportionality K is close to one.

When using the simplest theory one arrives at $K \approx 1.6$.

At room temperature, the highest known value of c is 1700 m/s for hydrogen for which M = 2. It becomes smaller for heavier molecules, for example, for the quicksilver vapors (M = 200.6) the c is only 170 m/s.

In the case of larger molecules, it is also necessary to consider the rotational motion of the molecules, inclusive of interatomic vibrations. Should the absolute value of c drop to zero, then the absolute temperature is also zero and no lower temperature can exist.

2.1.2 Pressure

The resulting force on a unit area of a limiting wall, i.e., the **pressure** is obtained by the integration of the momentum of the vibrating molecules weighed with the frequency of their impacts on the unit area of the wall

$$p \sim \rho c^2 \tag{2.2}$$

It follows from there that

$$pV \sim nMc^2 \tag{2.3}$$

So that it is apparent that the relationship between the mean velocity of the random translation and the temperature is in an agreement with the state equation of ideal gases.

2.1.3 The free path of molecules

The probability of a collision of two molecules grows with the number n of molecules in volume V and with their characteristic dimension d_M. Therefore the relationship of the mean free path L_M with these quantities will be given by

$$L_M \sim \frac{V}{n\, d_M^2} \tag{2.4}$$

so that it is directly proportional to the volume taken by one molecule and indirectly proportional to the area of its projection onto a plane. With the use of the state equation of the ideal gas, Eq. (2.4) becomes

$$L_M \sim \frac{R\,T}{p\, d_M^2} = \frac{M}{\rho\, d_M^2} \tag{2.5}$$

The volume of a molecule of the ideal gas at room temperature and pressure will be

$$V/n = \left(22.4\, m^3/kmol\right)/\left(6.022 \times 10^{23}\, molecules/mol\right) = 37.2\, nm^3.$$

Taken that the typical size of a small molecule is $0.4\, nm$, its free path will be around $50\, nm$.

2.2 Transport phenomena in gases

The moving molecules, while colliding with each other, equalize mutually their mechanical energy, meaning that they even out their respective speeds by **momentum transfer**. Thus in a continuum, this appears to be the result of the action of **viscosity**. Simultaneously, the intensity of their random motions is being equalized and that results in equalizing their **temperatures**. In a continuum, this will be seen as **heat**

transfer due to the **thermal conductivity**. Finally, the randomly moving molecules will exchange their respective random position, which means that, in a multicomponent system, this leads to the equalization of local **concentrations**. On a macroscale, this manifests itself as the **mass transfer** due to **diffusivity**.

2.2.1 Transport of momentum in gases

Let us consider a mass ($\rho L_M S$) closed between two planes having an area S separated by a distance L_M. An element of this mass will move from one plane to the other on average in time proportional to the ratio (L_M/c). If the difference of the mean velocities of these two planes equal to ($L_M\, du/dy$) then the **intensity of the flow of momentum**, that is the momentum transferred during the time period (L_M/c) by the area S

$$-\tau \sim \frac{(\rho\, L_M\, S)}{S\,(L_M/c)}\left(L_M\frac{du}{dy}\right) \tag{2.6}$$

In a continuum, the intensity of the flow of momentum is equal to the tangential stress τ and the ratio

$$\mu = \frac{\tau}{\left(\dfrac{du}{dy}\right)} \tag{2.7}$$

is called the **dynamic viscosity**.

By applying the model of momentum transfer by oscillating molecules, we shall get a proportionality

$$\mu \sim \rho L_M c \tag{2.8}$$

The **kinematic viscosity** defined as

$$\nu \equiv \frac{\mu}{\rho} \tag{2.9}$$

is the measure of the rate with which the speed changes propagate in a continuum.

By further consideration, it is possible to find the coefficient of proportionality to be

$$\nu \approx 0.499 c L_M \tag{2.10}$$

2.2.2 Heat transfer by conduction in gases

By analogy, should the temperature difference between two planes separated by a distance L_M be $(L_M\, dT/dy)$, then the **intensity of heat flow**, that is the heat energy q transferred in gases over a time period (L_M/c) across an area S is

$$q \sim \frac{(\rho\, c_P L_M\, S)}{S\,(L_M/c)} \left(L_M \frac{\mathrm{d}T}{\mathrm{d}y} \right) \tag{2.11}$$

The intensity of heat flow in a continuum will be proportional to the temperature gradient and the ratio

$$k = \frac{q}{\left(\dfrac{dT}{dy} \right)} \tag{2.12}$$

is called the **thermal conductivity**. Consequently, for gases it will be

$$k \sim \rho c_P L_M c \tag{2.13}$$

The speed with which the temperature differences propagate in a continuum, the so-called **thermal diffusivity**, will be given by

$$a \equiv \frac{k}{\rho\, c_P} \tag{2.14}$$

By applying a simplified theory it follows that for gases

$$a \approx 1.26 c L_M \tag{2.15}$$

2.2.3 Mass transfer in gases by diffusion

By analogy, should the concentration difference between two planes separated from each other by a distance L_M be $(L_M\, dC/dy)$, then the **intensity of mass flow**, that is the flow j of the component in question over a time period (L_M/c) across an area S is

$$j \sim \frac{(\rho\, C\, L_M\, S)}{S\,(L_M/c)} \left(L_M \frac{dC}{dy} \right) \tag{2.16}$$

The speed with which the molecules of a given component move in a stationary medium is expressed by a **diffusion coefficient** (**diffusivity**) D. For a gas mixture made up of molecules of similar sizes the D is approximately

$$D \approx 0.599 c L_M \tag{2.17}$$

The transport of mass by diffusion depends not only on the properties of the medium in which the diffusion takes place but also on the properties of the molecule moving through the medium. This is particularly striking in the case of liquids and solids in which the diffusion coefficients are lower than in gases by several orders of magnitude.

2.2.4 Prandtl and Schmidt numbers in gases

According to Eqs. (2.10), (2.15), and (2.17), the rates with which the speeds, temperatures, and concentrations, characterized by the coefficients ν, a, and D, even out, are very close. Therefore the interpretation of the transfers of momentum, heat, and mass transfer results in equations of similar type. In other words, these processes are analogous. The fact that the estimates of the numerical values of the transfer coefficients slightly differ from each other, results from the different averaging of the random motion of molecules for the three mechanisms. The ratios of the transport coefficients for heat and mass transfer known as the Prandtl number and the Schmidt number respectively, that for gases and vapors are close to one, can be used to assess the extent of the analogy. Thus the **Prandtl number** is

$$\mathrm{Pr} \equiv \frac{\nu}{a} = \frac{\mu c_P}{k} \tag{2.18}$$

And the **Schmidt number** is

$$Sc \equiv \frac{\nu}{D} = \frac{\mu}{\rho D} \tag{2.19}$$

For the air under room temperature and standard pressure, there is $\nu = 15.3 \times 10^{-6}$ m^2/s, $a = 21.5 \times 10^{-6}$ m^2/s, $D = 24.9 \times 10^{-6}$ m^2/s, Therefore $Pr = 0.71$ and $Sc = 0.61$.

2.2.4.1 The dependence of the transport coefficients of gases on temperature and pressure

The rate of signal propagation in a medium is, in general, proportional to the product of the mean velocity c of the random motion and the mean free path L_M of the molecules. When using the above relationships between 2.1 and 2.5 one gets

$$c\,L_M \sim \frac{\sqrt{R\,T\,M}}{\rho\, d_M^2} \tag{2.20}$$

It follows from this relationship that neither the dynamic viscosity μ nor the thermal conductivity λ depends on density and, consequently on pressure, which is in good agreement with experimental measurements. However, both these quantities just as the diffusivity D in gases grow with growing temperature.

There are, however, limits beyond which real gases cannot be approximated by the model of an ideal gas:

- At very low pressures the number of molecules in a given volume could be so small so that the free paths of the molecules could be larger than the characteristic dimension of the volume so that the laws of statistics no longer apply. Such a behavior is known as the **Knudsen's flow**.
- At very high pressures that are close to the critical pressure and higher, the molecules are so densely packed so that behavior of the gas comes close to that of a **supercritical fluid**; then the transport behavior is governed by the rules for liquids (supercritical fluids are remarkably good solvents).
- At the boundary between a high-pressure zone and a low-pressure zone, the equalization of the pressure could attain a speed significantly higher than the molecular motion due to their temperature. The velocity of the motion is then limited by the speed of sound and the step-like pressure change is known as the **shockwave**.
- At a temperature higher than 5000°C, gases become **plasma** when gas molecules disintegrate in individual atoms. At temperature still higher, the electrons are released from their paths and what is left are the cations with one or more positive charges. At temperatures higher than 10,000°C the transfer coefficients are significantly reduced.

2.2.5 The Knudsen flow

In very small spaces or at very low gas pressures, the characteristic dimension of the space becomes comparable, or even smaller than the free paths of the gas molecules Lm. A measure of this situation is the **Knudsen number** defined as

$$Kn \equiv \frac{L_M}{L} \tag{2.21}$$

For Knudsen numbers equal to one or higher, the system does not contain statistically significant number of molecules so that the molecules collide more often with the boundaries of the space rather than with themselves. The behavior of such a system differs substantially from what one would expect from the extrapolation of the behavior in a continuum. At room temperatures and atmospheric pressures, the mean free path of the molecule is around 50 nm so that one could expect the Knudsen flow to occur only in porous nanomaterials used for adsorption or for heterogeneous catalysis or when working under the condition of high vacuum.

At low pressures below 100 Pa, the free path of molecules extends over 50 μm so that the Knudsen flow can be observed in microreactors. An interesting application is the molecular vacuum pump (Fig. 2.1). In this device, a fast-rotating cylinder drives molecules of a gas in the direction of the rotation toward the exhaust. The narrow gap between the rotor and the stator would not allow the molecules to penetrate back to the suction opening.

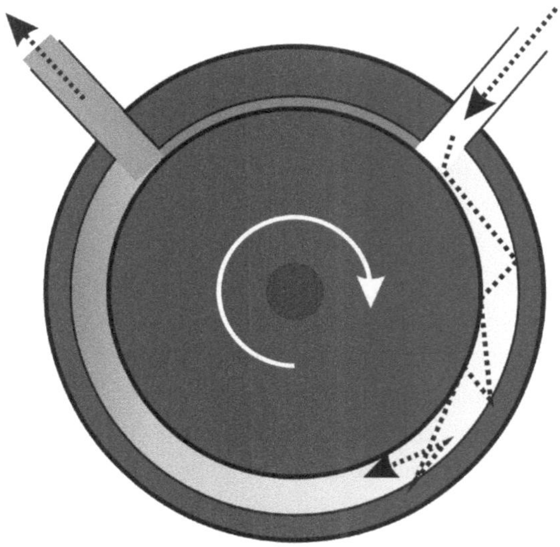

Fig. 2.1 Molecular vacuum pump.

2.2.6 Momentum transport at high-pressure gradients and at its sudden changes

Another limiting case for which the regular conclusions for transport phenomena no longer apply are systems displaying high-pressure gradients. Examples of such situations include a case with fast-moving boundary, flow from jets, and explosions. The common characteristic of all these situations is that the ratio of their speed to the speed of sound, the so-called **Mach number**, is higher than one

$$\mathcal{M} \equiv \frac{U}{U_{sound}} \sim \frac{U}{c} \tag{2.22}$$

so that the process takes place at a **supersonic speed**. When taking into account that the sound moves faster at a higher pressure, the information moving from the source accumulates in one line known as the shock wave. The passage of the **shock wave** past an observer is accompanied by a strong acoustic, aerodynamic boom. In small spaces, one observes the formation of small local shock waves that is perceived as hum, whistling, rustle or hissing or the sound of crackling bubbles.

Adiabatic changes taking place in the vicinity of a shock wave are dealt with by the thermodynamics of explosions.

The behavior of systems at subsonic and supersonic speeds is the key subject of aerodynamic studies, particularly when designing planes and missiles but it will be also encountered while designing turbines and jets.

2.3 Momentum transport in solids and liquids

While in gases at room temperature and atmospheric pressure, where the molecules and atoms have the size of 0.1–0.5 nm and where they are spaced approximately 3.4 nm, the particles in liquids are substantially larger and far from being spherical. In addition, they are densely packed and they are subject to a significant attraction force. In solids, these attraction forces make the particles immobile. In crystals, the particles assume spatially regular positions. In liquids, the spacing between particles while limited, it is adjustable. It is obvious; however, that the simple theories of transport phenomena derived for gases are unlikely to hold for gases and vapors.

2.3.1 Tensile test

A simple test in which a piece of material is subjected to one-dimensional stress is used to study changes caused by a force acting on a solid material.

Such a standard tensile test uses a test body (usually a rod or a prism, Fig. 2.2) of a known cross section S_{x0} subjected to a **very slow extension** in the direction x while measuring the applied force F_x. Usually, the stress sigma is plotted against the relative extension epsilon. The **relative extension (strain)** is then defined as the ratio

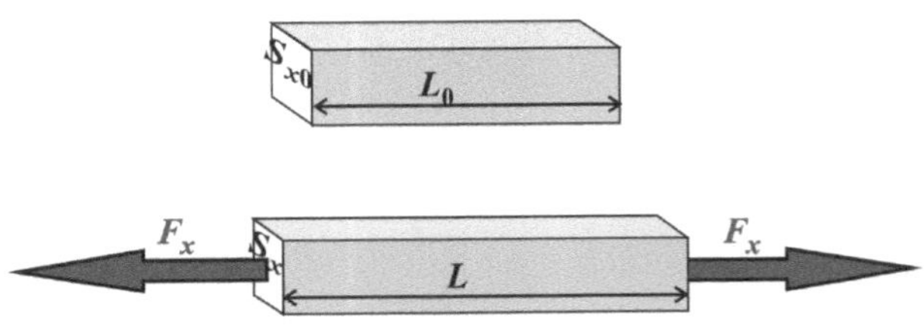

Fig. 2.2 Tensile test.

$$\varepsilon \equiv \varepsilon_{xx} = \frac{L - L_0}{L_0} \tag{2.23}$$

where L_0 is the original length of the test body in the x direction and L is the length after extension.

Another variable measured is the **contractual tensile stress** σ_0

$$\sigma_0 \equiv \frac{F_x}{S_{x0}} \tag{2.24}$$

It is defined as the force in the x direction related to the original cross section of the test body. A typical relationship is plotted in Fig. 2.3.

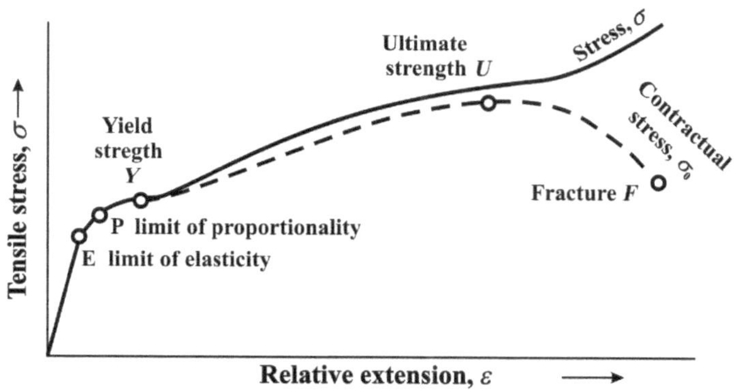

Fig. 2.3 Tensile test record: ____ actual tensile stress and - - - - - contractual tensile stress.

The actual tensile stress, σ is higher than the contractual one because it has to respect the narrowing of the cross section of the test body

$$\sigma \equiv \sigma_{xx} = \frac{F_x}{S_x} \tag{2.25}$$

The upper limit of the validity of **Hooke's law**

$$\varepsilon_{xx} = \frac{\sigma_{xx}}{E} \tag{2.26}$$

is given by the limit of proportionality. The elastic reversible deformation takes place up to this limit $0 < \sigma_{xx} < \sigma_U$.

To some extent, Hooke's law applies also to the calculation of the pressure stress, with the reversed—$\sigma_U < \sigma_{xx} < 0$, where $\varepsilon_{xx} < 0$.

Alongside the extension of a body the narrowing of its cross section takes place

$$\varepsilon_{yy} = \frac{\sqrt{L_0} - \sqrt{L}}{\sqrt{L_0}} \tag{2.27}$$

In the region of elastic deformation, every material is characterized by the ratio

$$\mu_p = \frac{\varepsilon_{yy}}{\varepsilon_{xx}} \tag{2.28}$$

known as the **Poisson number**.

The value of the **module of elasticity** E as defined by Eq. (2.26) for different classes of materials may vary by several orders of magnitude. For illustration, typical values are given in Table 2.1.

Table 2.1 Typical results of slow tensile test at room temperature.

Material	Module of elasticity, *E* (GPa)	Limit of elasticity (MPa)	Ultimate strength (MPa)
Carbon fiber	1000		10000–100000
Metals and alloys	50–500	50–2000	200–5000
Construction steel	200	250	400–550
Glass and ceramics	20–100	1000–1500	1500–3000
Wood parallel to grain	4–20		
Plastic	0.1–5	10–50	20–100
Rubber	0.001–1		10–20

To get an idea about the meaning of the above numbers let us consider a wire having a cross section of 1 mm^2. If this wire is loaded with a weight of 50 kg in a normal gravitation field without being torn (although it might be extended), then its strength yield point is 500 MPa.

2.3.2 The effect of force at fast straining

Normally, the tensile test is carried out at very small speeds. Solid materials usually have a significant shape memory, which, for liquids, is negligibly small. The application of tensile stress for a short, finite time may result in a smaller deformation than it would occur after a slow application of the stress for a long time. However, the length of the application of the tensile stress below the limit of proportionality has essentially no effect. After, however, the limit of proportionality is exceeded, the yield stress is soon reached when the particles of the material begin to regroup so that irreversible plastic reforming occurs; in this case, the duration of applying the stress plays a significant role. Schematically, this is shown in Fig. 2.4. Unlike the plot in Fig. 2.3, the axis is switched because it is more fitting to consider the stress to be the independent variable.

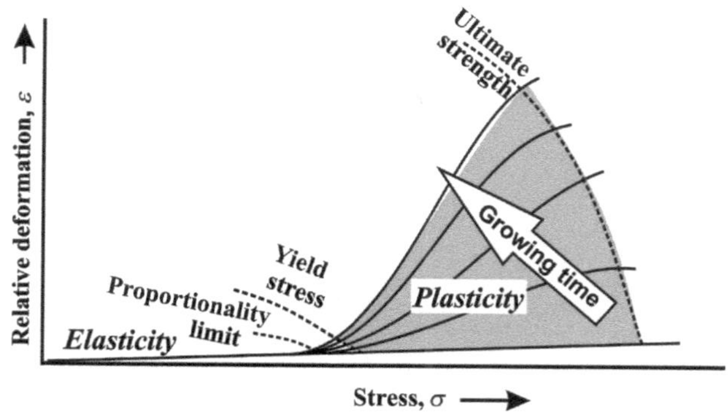

Fig. 2.4 The effect of fast straining.

The rate of spreading of the elastic deformation after a sudden change of applied stress can, at the most, reach the speed of sound. For a casual observer, this might appear as "infinitely fast."

A good example of a very fast elastic deformation and its response is the impact of billiard balls with each other and with the walls (impact of elastic bodies).

2.3.3 Shear stress

Apart from applying a tensile and pressure stress, the material can also be strained by a shear stress, a phenomenon of particular significance in the case of transport phenomena. This is shown schematically in Fig. 2.5. Qualitatively, the shear strain dependence on the shear stress is analogous to the tensile stress as shown in Fig. 2.4 but quantitatively there is a shift.

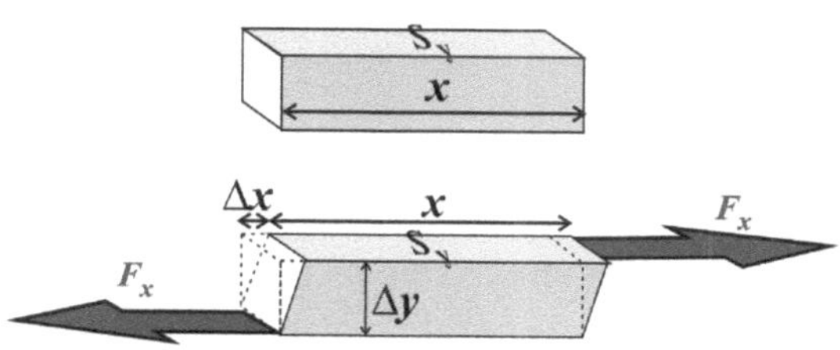

Fig. 2.5 Strain test.

Also on the shear stress $\sigma_{yx} \equiv F_x/S_y$ curve, there will be a limit of proportionality, a yield point, and a strength limit. In the elastic region at small σ below the limit of proportionality, there will be a direct proportionality of relative shear deformation $\varepsilon_{yx} \equiv \Delta x / \Delta y$ with the applied shear stress, σ_{yx} analogous with Hooke's law.

$$\varepsilon_{yx} = \frac{\sigma_{yx}}{G} \tag{2.29}$$

where the module of elasticity of the shear strain G is somewhat smaller than one-half of E.

$$G = \frac{E}{2\left(1 + \mu_p\right)} \tag{2.30}$$

The value of the dimensionless Poisson number for the usual construction metals is around 0.3.

The experimental determination of the shear stress in usual construction materials is difficult. It is, however, relatively easy for materials that can tolerate large shear strains, that is for strongly elastic or plastic materials. Most often, these will be semiliquid materials representing a transition between liquids and solids such as polymer melts and solutions, suspensions and slurries, emulsions and pastes etc. In liquids, the dependence of both the stress as well as the strain on time is roughly linear.

2.3.4 Shear strain of fluids

Elastic behavior in fluids (both liquids and gases) is essentially nonexistent even after applying small stresses. Fluids exhibit usually only plastic deformations that increase with progressing time. When dealing with liquids, it is useful to define the **rate of shear deformation**

$$\gamma \equiv \frac{d\varepsilon}{dt} \tag{2.31}$$

For which a unique assignment exists

$$\gamma = f(\sigma) \tag{2.32}$$

Only in the case of a **simple shear strain** (Fig. 2.5) it is possible, with the use of Eq. (3.8), to equate the rate of shear deformation with the velocity gradient

$$\gamma \equiv \frac{d\varepsilon_{yx}}{dt} = \frac{du_x}{dy} \tag{2.33}$$

However, for more complex geometries, the rate of shear deformation will differ from the velocity gradient. (More on this will be discussed in the Section 8.2.)

The rate of shear deformation as given in Eq. (2.32) will be thus related to a particular component of the tensile stress

$$\gamma = f\left(\tau_{yx}\right) \tag{2.34}$$

For most ordinary liquids, so-called Newtonian liquids, the function (2.32) is linear and it is termed as Newton's law of viscosity

$$\tau_{yx} = \mu \frac{du_x}{dy} \tag{2.35}$$

where the coefficient of proportionality is, according to the definition (1.8) the dynamic viscosity μ having the dimension of $\mathrm{Pa\,s} \equiv \mathrm{kg\,m^{-1}\,s^{-1}}$.

2.4 Rheology

Experimental methods used to measure the dynamic viscosity of Newtonian liquids are called the **viscometry** and the instruments used for the measurement, be it those commercially available or those home made, are called **viscometers**. When used for non-Newtonian liquids, they can determine small sections of the relationship between shear stress and shear rate.

In general, science that deals with the deformation of materials and their changes with time (that is with the flow of liquids) being dependent on the applied stress is called **rheology**. The related discipline dealing with experimental determination of these relationships is **rheometry**.

As shown above, deformation of a material depends not only on the stress acting on it but also on the duration of time over which the stress is applied. Mathematical modeling of these relationships results in the complicated differential equation the realm of which is known as the **mechanics of continua**.

2.4.1 Deformation of material by a step change

In ordinary liquids at rest, the deformation stress (except for the normal component of pressure) is equal to zero. When a step deformation is applied to a Newtonian liquid, the stress will return to zero in a very short time. The length of the return is controlled purely by physical forces such as the inertia.

In the case of purely elastic materials, the stress after a deformation remains for a while at a constant value determined by the Hooke's Law. However, should the deformation be sufficiently high, some extent of plasticity might occur and the flow and the stress slowly subside. This is called the **stress relaxation**. This phenomenon is best described by the **relaxation time** that is defined as the time during which the stress drops *e*-times (where "*e*" is the natural logarithm of the stress in chosen units dropping by one). A Deborah number *De* is used as the measure of the importance of the relaxation time

$$De \equiv \frac{\text{Relaxation time}}{\text{observation time}} \tag{2.36}$$

Should the *De* for a certain material approach zero then this material will display a liquid-like behavior. If, on the other hand, the Deborah number approaches infinity, then the material in question will be a perfect solid elastic. Any material having the *De* a finite value between zero and infinity will be a viscoelastic material.

The name of this dimensionless number is derived from the name of a Jewish prophetess who might have said "The Mountains shook before the Lord..." (Old Testament/The Book of Judges 5:5) meaning that for a long-range observer even the mountains grow fast.

2.4.2 Cyclic deformation of materials

A rheometric experiment subjecting the material to a cyclic harmonic strain

$$\varepsilon = \varepsilon_0 \sin(\omega t) \tag{2.37}$$

can be used to classify the tested material since the stress response will depend on both the frequency as well as on the amplitude of the stress imposed on the material. From this information, it can be then gleaned, which properties of the material are dominant and which can be neglected.

- In the case of **elastic materials and small amplitudes**, the stress remains in the limits of the Hooke's law so that the stress response will again be a harmonic function,

$$\sigma \equiv E\varepsilon = E\varepsilon_0 \sin(\omega t) \tag{2.38}$$

and in phase with the deformation.

o The response of liquids will also be a harmonic function and the stress amplitude will be proportional to the derivative of the deformation

$$\sigma \equiv \mu \frac{d\varepsilon}{dt} = \mu\, \omega\, \varepsilon_0 \cos(\omega t) = \mu\, \omega\, \varepsilon_0 \sin\left(\frac{\pi}{2} + \omega t\right) \tag{2.39}$$

The lowest response will be then in the instance of the largest deformation when the speed of the motion comes to zero and the motion changes the direction. Thus the response will be shifted by one-quarter of the cycle that is by 90 degree.

o Both the viscosity as well as the elasticity manifests themselves in the simplest **viscoelastic materials** and the response to a harmonic oscillatory deformation is a harmonic function shifted by an angle. The relaxation time can be derived from the **phase shift**.

o After the yield point is exceeded, both the **elastoplastic material** as well as the **viscoplastic liquid** will display some plasticity. Under these conditions, a hysteresis appears and the response curve loses its sinusoidal shape.

Whenever a material is tested by a cyclic harmonic deformation, it is important to choose the amplitude and the frequency in such a range that corresponds to the condition under which the material is processed and used. In some cases, only strong short impacts are of interest while in other cases, one is focused on a long term, mild straining.

The responses of rheologically complex materials can be simulated by the behavior of simple mechanical elements. So the elasticity can be represented by a spring, viscosity by a damper, and plasticity by a brake. Their behavior can be described by simple equations. Serial and parallel combination of these three mechanical elements can create an analogous behavior of complex rheological materials.

Another analogy can be created by combining electrical circuits with an impedance that can be obtained by branched series of resistances, condensers, and spools.

2.4.3 Structural rheology

Although classical physics treats all materials as if they were continua, it is not possible to ignore the fact that they are not. A full understanding of their behavior is not possible without taking into account their internal structure. The specific section of rheology dealing with the relation between the structure of materials and their behavior under deformation is called the **structural rheology**. This discipline sees materials as being ensembles of mutually affecting particles. These particles could be atoms arranged in a metal lattice, ions in crystals, molecules or their aggregates (macromolecules), or large domains of particulates ranging in size from nanometric to micrometric to macroscopic dimensions. The distances of individual atoms in molecules are around 0.1 nm that is 10^{-10} m. The same applies to individual molecules as well as to atoms in crystals or in metallic lattices. At these distances, whether chemical, ionic or metallic bonds are present or not, the attractive forces between individual bodies are equally significant as the repulsive forces and, therefore the materials exist in a **solid state**.

2.5 What happens while straining condensed materials?

While applying relatively **small deformations** on condensed materials (that is other than gases), the particles just change their mutual separations without altering their configuration. Up to a certain point (the limit of proportionality), the deflection is directly proportional to the acting force. Thus these changes are **elastic**. They are also reversible: when the stress elapses the material will return in the original state. This generally applies to all types of strain, be it traction, pressure, bend, shear, or torsion. When the **limit of elasticity** is exceeded then **plastic slippage** begins to occur whereby the configuration of particles keeps on changing without the attraction among them being overcome. The ranges of these regions depend significantly on temperature, as shown in Fig. 2.6.

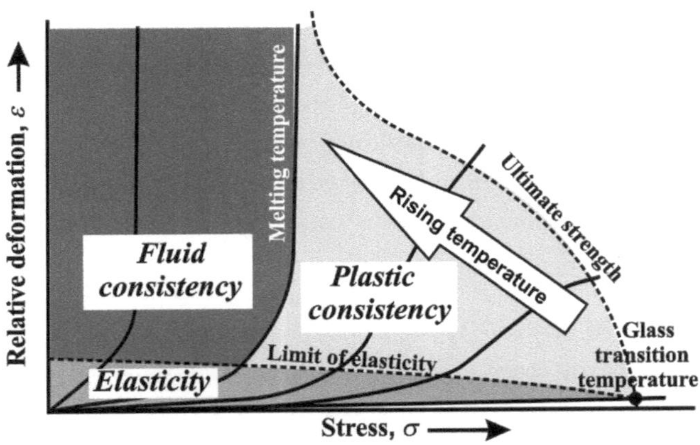

Fig. 2.6 Straining at different temperatures.

In liquids, that is above the melting point, the influence of elasticity is usually negligible compared with the increasing deformation due to the applied stress. Individual layers in the **metal** grids formed by individual atoms, begin to slide while these shifts usually begin and stop at a fault in the original grid. Materials are often formed by self-contained blocks, each of them containing a well-organized lattice that has the form of a grain of a micrometric dimension. Then the faults usually occur at the boundaries of these grains that will undergo deformation and, finally, a breakup into smaller grains.

In the case of **macromolecular materials** formed by balls of long macromolecules, these balls will get elongated and the macromolecules will get somewhat straightened and rearranged. These changes lead to an orientation of the polymeric material, initially effortless and then gradual. The size of the range and the rate of plastic deformation rise significantly with temperature (see Fig. 2.6). The transition to a liquid state (that is the melting temperature) is not very pronounced for materials having high viscosity of the melt. This type of behavior is exemplified by the melting of **glass** at increased temperatures.

In general, plastic deformation is rather slow so that the final result depends both on the duration of the stress applied. Plastic deformation occurring over a period of days or even years is known as **plastic flow** of materials or **creep**. A certain degree of deformation leading to a certain degree of particle orientation results in the **reinforcement** of the material. The plastic deformation stop at this stress.

Should the force, and with it the stress, during the elastic deformation cease to act, then what follows is the **relaxation** whereby the body returns to the original shape due to release of the reversible elastic energy. Should, on the other hand, the stress continue to grow then the material can no longer resist to the outside force; its **strength limit** is exceeded and the material loses its cohesion (**breakage, fracture, crushing**). The size of the plastic region decreases with decreasing temperature. If the temperature drops below the **temperature of vitrification** then the material becomes **fragile**: the limit of elasticity and the tensile strength almost coincide.

A pronounced elasticity can be observed even in coarse macroscopic **composites**, such as foams, mosses, and fibrous materials (polyurethane foam, felt, tissues, hey).

In **granular materials** (powders and grains), the attractive force between particles is so small that the traction test cannot be performed at all since the traction strength is negligibly small compared with the pressure strength.

The yield strength of concrete in traction is 0.5–2 MPa while in pressure is 5–30 MPa.

2.5.1 The transition from the solid state to liquids as seen by the mechanics

Although the response to strain is similar for all solid materials (be it at different levels of magnitude and in different ranges of variables), there is no equation that is used to predict the behavior of solids as is the case with vapors and gases for which the state equation of an ideal gas predicts the behavior reasonably well, even if the gas molecules were of different sizes. The only systematic rule applicable to solid materials states that at higher temperatures the atoms and the molecules, including their clusters, are more mobile (within the confined boundaries), which leads to the increase of the repulsive forces between the particles and, consequently, to easier workability. Schematically, this is shown in Fig. 2.7.

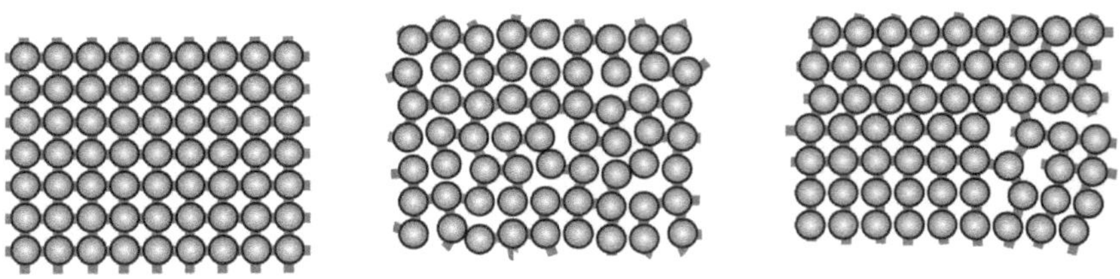

Fig. 2.7 Structure of a solid material: uniform structure, relaxed structure with dislocations sliding and flow on straining.

In summary, it can be concluded that the influence of stresses on solid materials and their internal structure at a given chemical composition, depends not only on the instantaneous stress but also on its history of changes, as well as on the history of the temperature changes to which the solid material was subjected. An attempt to derive an equation with a predictive capability for solid materials would result in complex integral relationships.

A sharp temperature boundary called the melting point exists for all ordinary pure low-molecular solids that separate the liquid and the plastic regions. Above this temperature, the elasticity is almost nonexistent while the plasticity is very limited below this temperature. There might exist, however, below the melting point other phase transitions separating different modes from crystalline corresponding to different temperatures or pressures.

For example ice can exist at more than ten crystalline modes at different temperatures and pressures. Further, a thin layer of amorphous water exists at the phase boundary between the ice and gas at temperatures higher than $-33°C$ that is not bound by hydrogen bonds as it is in the crystalline ice. This quasi-liquid layer, that is responsible for the lubricity of ice, reaches a thickness of several tens of nanometers at $-1°C$.

In general, it has to be recognized that miscellaneous attractive and repulsive forces of different sizes and modes exist that keep the particles at specific spacing. While all solid materials behave in a similar fashion, their quantitative measures will differ as can be seen from Table 4.1. Schematically, a set of particles is depicted as a set of arranged spheres in Fig 2.7.

2.5.2 Crystals

A slip in a perfectly spatially arranged crystal grid usually does not occur. Rather, the material remains in the given configuration with only an infinitesimally small reversible deformation until it reaches the yield strength when the applied force exceeds the attractive force among atoms that holds them together by a chemical or an ionic bond. However, there are cases when a slippage can occur if the strength is applied in one specific direction.

While the carbon in the diamond crystalline mode has a very firm crystalline structure making a slip impossible, the same carbon in the graphite mode, formed by flat hexagonal scales, will facilitate slippage freely. Another carbon mode is fibers formed by carbon nanotubes that have infinitesimally small elasticity, negligible plasticity but an extremely high tensile strength.

If the crystal grid of a material is formed by individual low-molecular ions then the material will usually be characterized by a sharp transition to a molten state at the temperature of the melting point (provided that the melting is not preceded by a chemical destruction). This is because the rising temperature relaxes the attractive forces among the molecules. Then only a small outside force will suffice to bring about jumps of molecular chains along with the shifting of local irregularities—vacancies (cavities). If, however, the crystal is made of larger objects connected together by covalent bonds (e.g., inorganic macromolecules such as the quartz), gradual softening can occur (glass) and the transition to a molten state is not sharp.

2.5.3 Metals

The slip (malleability and ductility) can be evident in pure metals already at the atomic level (at 0.1 nm). However, when a metal crystallizes from a melt, a regular crystalline grid will not be always formed. This is true, particularly for alloys. Instead, regularly arranged clusters of metal crystals will form grains up to several micrometers large that contain up to billions of atoms. Each of these grains might behave as a malleable metal. However, each grain has a differently oriented structure and any dislocation will stop in such an irregular structure at the grain's boundary.

In the case of steel (Fig. 2.8), the grains are formed by pliable, malleable austenite that has a regular structure of layers while the space between the grains is filled by martenzit and by other hard, brittle components.

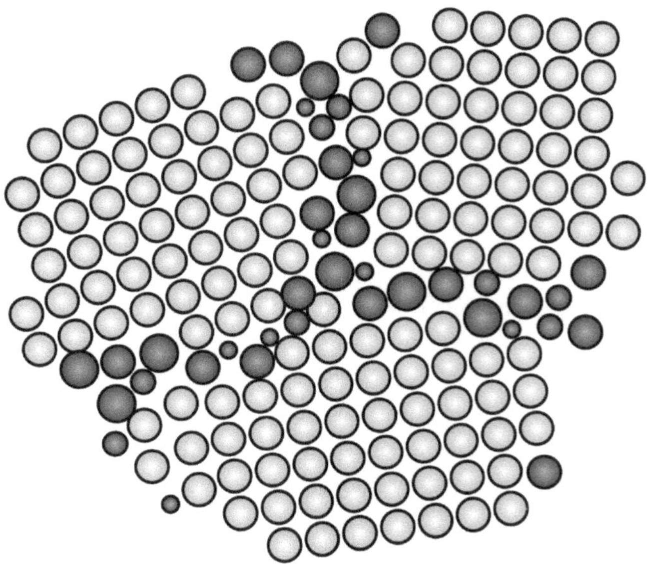

Fig. 2.8 Ductile grains and amorphous interfaces.

To increase the firmness of such material, it is advantageous to decrease the size of the grains by forming, for example, by forging or rolling. Therefore forming not only changes the shape of the material but also its internal structure (see Fig. 2.9).

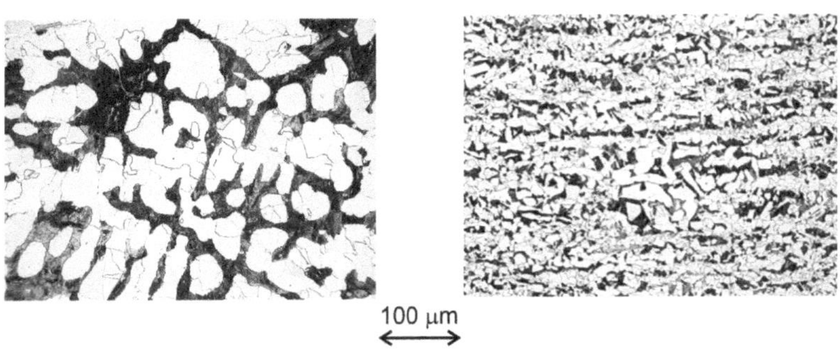

Fig. 2.9 Structure of a steel cast: original crystalline structure and after the hot rolling.

By plastic deformation, the grains are split and elongated and this results in a more regular internal arrangement of the material and, therefore, to a reduction of dislocations. Any straining of such metal is spread into a larger number of smaller dislocations and, thus to the **strengthening** of the material compared to the strength of the original cast that originated just by crystallization.

A coarser structure can be brought about by the addition of foreign components to the alloys. A typical example is the behavior of cast iron with a high content of carbon and silica.

Higher temperatures will result in an increased plastic deformation of metals resulting in the increase of their integrity. This is because at higher temperatures the bonds between atoms and between larger particles will loosen leading to the formation of gaps into which the particles could move when an outside force is applied. Therefore most of the forming processes are carried out at increased temperatures.

The long-term straining of a solid material close to its yield stress will bring about a very slow flow called the **creep**.

A good example of creep is the slow motion of glaciers moving through valleys to lower positions. Often large cracks will appear perpendicularly to the direction of flow.

In general, the final elasticity, plasticity, and the strength of metals depend not only on the elemental composition but also in a complicated fashion, on the history of the heat processing (annealing, tempering, quenching) as well as on the history of forming during the previous processing. This is because the histories determine not only the size but also the crystalline structure of the grains of the metal as well as of the admixtures. To increase the usefulness of the final metal products (their hardness or their resistance to corrosion), they can be doped by **other admixtures**.

In the old days, the blacksmiths worked with a relatively soft iron. Therefore they had to treat the edges of tools and the blades of weapons with carbon to create a hard coating on them. On the other hand, too much carbon is dissolved in the pig iron flowing out of the blast furnace. Therefore this carbon must be subsequently removed by complex steel-forming technologies.

By melting a metal (that is by transforming it into a liquid form that usually has very low viscosity) the cast will lose all acquired properties so that metals can be recycled without difficulties.

2.5.4 Polymers

Polymers are formed by chemical reactions that bind individual molecules in linear or branched **chains** or in spatial **networks**. Polymers formed by short linear chains will become liquid at higher temperatures. They might also dissolve in suitable solvents. During both such liquefactions, the balls of polymers will straighten up, either reversibly (elastic) or irreversibly (plastic). This manifests itself as a reduction of viscosity with increasing deformation rates. This phenomenon is known as the viscoelasticity and pseudoplasticity of polymer solutions and melts. Such liquids are known as the **non-Newtonian liquids** because they do not follow the Newton's law of viscosity that states that the viscosity is independent of the deformation rate.

By choosing a specific method of polymerization it is possible to prepare plastics formed by linear polymer macromolecules into orderly arranged pliable structures called **crystalline polymers** or into randomly arranged **amorphous polymers**. Additionally, the arrangement of the macromolecules can be affected by a controlled deformation of the polymeric material, the so-called **orientation of the linear chains** (Fig. 2.10).

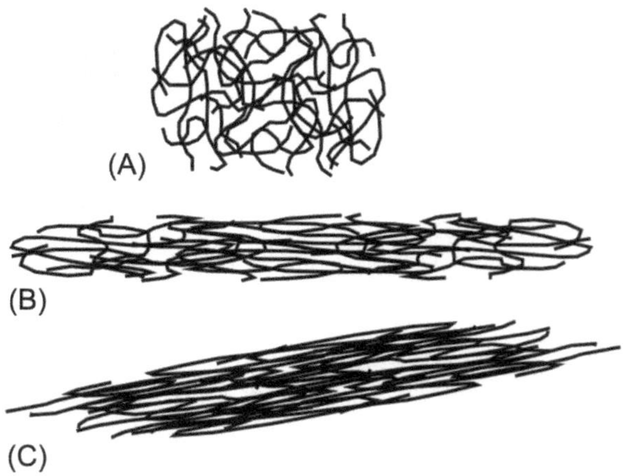

Fig. 2.10 The reinforcement of a polymer fiber ball by orientation: amorphous structure, after the prolongation, and after the shear strain.

To some extent, this is reminiscent of the manufacture of textile yarn from a staple and the subsequent twisting of it into a thread or a rope.

Mutual mobility of polymer molecules decreases significantly with decreasing temperature. Below the so-called **temperature of vitrification**, the polymer becomes brittle so that a fracture becomes more common than a plastic deformation. Above the temperature of vitrification, the polymer nets tolerate only elastic deformation. For thin networks, such as rubbers, this reversible deformation might be very significant, tens or even hundreds percents. The volume and the mechanical strength of them can be significantly modified if they are swollen by suitable low-molecular solvents.

Linear and weekly cross-linked polymers can be dissolved in solvents to form solutions that preserve a part of the elastic properties.

The mechanical forming of polymers can be more easily done when they are first plasticized at higher temperature with the subsequent hardening of the product by suitable chemical agents (the making of reactoplasts or the vulcanization of rubber). Materials that are cross-linked in this way will preserve the region of elasticity to some extent. However, should the stress exceed the limit of elasticity, they will disintegrate without passing the region of plasticity and flow. Cross-linked materials cannot be liquefied again by just increasing the temperature without using some kind of chemical destruction.

When processing materials that melt easily, the product can be formed straight from the melt. The process is then completed by the reduction of the temperature while maintaining the pressure to compensate for a certain shrinkage that always occurs at solidification. However, not even the linear polymers can be returned to the original state by melting because of the repeated reforming results in gradual mechanical destruction of the macromolecular chains.

2.5.5 Viscosity of liquids

Just as the structures of solid materials differ from one to another, it cannot be expected that the liquids formed by their melting, would display uniform behavior at deformation and flow. Despite repeated attempts to find some predictive formulas for the estimation of the viscosity of liquids, most of the molecular models of flow are usable only for limited classes of similar molecules.

The so-called Eyring hypothesis assumes that the regrouping of molecules in liquids can be visualized as mutual avoidance of molecules moving in more or less parallel layers. If this model is used for small, sphere-like molecules it will predict that

$$\mu = \frac{N\,h}{V_M} \exp\left(\frac{E_\mu}{R\,T}\right) \tag{2.40}$$

where V_M is the volume of 1 mol of the liquid and $N = 6.023 \times 10^{23}$ and $h = 6.626 \times 10^{-34}$ J s are the Avogadro and Planck constants, respectively. The creation of vacancies between the rows of molecules that facilitates the shifting of the layers, that is the flow, will increase exponentially with increasing temperature. This is expressed by the Arrhenius-type relationship where E_μ is the activation energy in the formula that relates the dependence of viscosity on temperature. As a rule, the viscosity of liquids drops with increasing temperature, which is the opposite case for gases and vapors the viscosity of which always grows with increasing temperature.

It can be assumed that the E_μ is proportional to the evaporation enthalpy $-H_{evap}$ at the regular boiling point. When combining this assumption with the **Pictet-Troutoun rule** from which it follows that the evaporation enthalpy is proportional to the boiling temperature and with the realization that the viscosity of organic liquids at the boiling point is usually 0.23 mPa s, one gets an approximate formula for the viscosity estimate

$$\mu = 0.23 \exp\left[3.8\left(\frac{T_B}{T} - 1\right)\right] \tag{2.41}$$

The viscosity of the liquids is not appreciably dependent on **pressure**.

The above formula fails for liquids in which there are additional bonds between the molecules, for example water and, among the organic liquids, alcohols for which the viscosity is several times higher than that predicted by Eq. (2.40). This formula also fails for liquids that have a very high boiling point such as glycerol, heavy oils, concentrated sugar solutions, and other liquids for which the temperature dependence of viscosity is truly dramatic. On the other side, there are also liquids for which the viscosity is only minimally dependent on temperature, such as certain silicon oils. Analogous relationships, based partly on theoretical considerations, partly on the correlation of experimental data, were also proposed for molten metals. Their dynamic viscosities are relatively low, typically between 1 mPa s for alkaline metals to 6 mPa s for uranium. Viscosities of molten metals drop only mildly with increasing temperatures. In view of their higher density, their kinematic viscosities are substantially lower than that of water.

Formulas to predict the viscosity of liquids are not very important because the values of viscosities for most liquids and of liquid mixtures are available in tabular form in the technical literature. (Some typical values are shown in Table 2.2.)

Further, devices for the measurement of the viscosity of liquids are available in most standard laboratories. Much more difficult, and costly, is the study of the flow of non-Newtonian liquids that display a pronounced nonlinearity between the stress and the shear rate and, additionally, clear signs of plasticity and elasticity. The more detailed treatment of these phenomena would, however, fall outside the scope of this text.

Table 2.2 Viscosiity of liquids (50°C).

	μ (mPa s)
Steam	0.0100
Air	0.0196
Methanol	0.396
Water	0.561
Transmission Oil	95
Glycerin	180
Honey 85%	1500
Silicon oil	19000
Mercury	1.4
Steel (1600°C)	2.6

2.5.6 Viscosity of liquid mixtures

Mixtures of **gases and vapors** are always homogeneous and their viscosities are determined by the viscosities of their components approximately in the ratio of their molar fractions.

Homogeneous liquid mixtures, including dissolved gases, salts, and other dissociated ionic matters at low concentration have a viscosity not too much different from the viscosity of the pure solvent. Concentrated solutions of larger non-dissociated or weakly dissociated molecules (e.g., starch, honey, glycerin but also the concentrated sulfuric acid) are usually highly viscous but their viscosity drops rapidly by the addition of the low viscous solvent. The speed with which the viscosity drops depends on the molar mass. A universal formula for the prediction of their flow behavior is not available but, in any event, it is usually much simpler to measure the viscosity than to determine the concentrations of the components. In reality, the experimental measurement of the viscosity of diluted polymer solution is used to determine the molar mass of the polymer.

In general, the flow behavior of **solutions and melts** of polymers made of linear or cross-linked macromolecules is very complex. In flow, the chains will straighten up and that causes the apparent viscosity of the mixture to drop with increasing shear stress. Furthermore, at very high shear stresses the long linear macromolecules will suppress the transition to the turbulent flow. Even at very low concentration of several hundreds of ppm, these macromolecules will cause the extension of the laminar flow region to very high Reynolds numbers.

This effect is utilized to reduce the flow resistance of submarines and torpedoes by injecting polymer solutions to the tip of these objects. A similar function is provided by the slime on the surface of fish and sea mammals.

Heterogeneous liquid mixtures are two-phase liquid mixtures made of by a homogeneous phase with a dispersed solid phase (**suspension)**, or an immiscible liquid (**emulsion**) or a gas (**foam**). Mixtures with miniscule dispersed particles (statistically small, that is in comparison with the size of the observed volume) behave as homogeneous liquid (e.g., milk). Small liquid particles maintain their spherical shape while mutually obstructing their own motion. The estimation of the viscosity mixtures containing a small fraction of dispersed particles serves the so-called Einstein's formula

$$\mu \approx \mu_0(1 + 2.5\varphi) \tag{2.42}$$

where φ is the fraction of the dispersed phase and μ_0 is the viscosity of the continuous phase. If the volume fraction of the dispersed phase is large then the particles will touch each other and the experimentally determined apparent viscosity will increase. The suspensions then will become pastes, the emulsions will become creams, and the bubbles in liquids will form solid foams. Usually, such materials will display a non-Newtonian behavior with the signs of plasticity or elasticity. Other irregular behavior will be observed even at low concentrations of particles if there is an attraction among particles as is the case with the plate-like particles of clay.

2.6 Heat transfer by conduction in liquids and solids

2.6.1 Thermal conductivity in liquids and solids

There is no simple molecular model to determine the thermal conductivity (definition 1.3) of the condensed phase. It can be expected that for both liquid as well as solid organic materials the thermal conductivity will be around $k \approx 0.2\,\mathrm{W\,m^{-1}\,K^{-1}}$, For water, ammonia and most inorganic materials it will be $k \approx 0.5\,\mathrm{W\,m^{-1}\,K^{-1}}$. The thermal conductivity of solids is somewhat higher than for liquids; for example ice will have $k \approx 2\,\mathrm{W\,m^{-1}\,K^{-1}}$. It will be even higher if these materials are in a crystalline mode because of the occurrence of free electrons. It will be higher still should the crystal structure resemble the regular structure of metals.

The thermal conductivity of both solid as well as molten metals will be of the order of several hundreds. Just as is the case with electrical conductivity, the thermal conductivity of pure metals is higher than that of alloys.

Numerical values of some common materials are given in Table 2.3.

As for composites, that are materials composed of numerous phases—macroscopic particles having varying conductivities, determine the conductivity of one or another phase. Composites containing a large fraction of gas

Table 2.3 Typical thermal properties of selected materials at room temperature.

	λ (W/(m K))	c_P (J/(kg K))	ρ (kg/m³)	a (10^{-6} m²/s)
Steam	0.02	2000	0.6	17
Air	0.025	1000	1.2	21
Polyethylene	0.4	2800	900	0.16
Polystyrene	0.2	1400	1030	0.14
Water	0.59	4200	1000	0.14
Glass	0.8	840	2700	0.35
Ice	2	2000	910	1.1
NaCl	5	870	2160	2.7
Fe	80	455	7870	22
Stainless steel	20	460	7700	5.6
Al	230	902	2710	94
Cu	370	386	8930	107
Ag	420	234	10500	170
Graphene	5000	700	2250	3200
He II (2K)	>10,000			

(having a high porosity), such as granular or fibrous materials, or foams with individual bubbles, or mosses with connected channels, the heat conductivities are much lower than in coherent materials. For this reason, the porous materials are used as heat insulators.

Thermal conductivity of solids can be conveniently measured by commercially available instruments. The case with liquids and gases is more complicated because the natural convection, that cannot be avoided, will distort the reading.

2.6.2 Heat capacity of liquids and solids

Heat capacity is defined as the amount of heat needed to increase the temperature of a unit of mass by 1°. The **calorimetry**, which is the experimental technique used to measure the heat capacity, is a well-developed method because this variable is frequently used in thermodynamics. The numerical values of heat capacities of pure materials, common mixtures, and technical materials have been extensively determined by measurement and they are published in tabular form. Alternatively, empirical formulas for their determination in a wide range of temperatures are available.

For example, the NIST (National Institute of Standards and Technology).

In engineering applications, the range of temperatures is rarely not large enough to permit, considering, the heat capacity to be a constant.

2.6.3 Thermal diffusivity and the Prandtl number in liquids and gases

The thermal diffusivity of gases was dealt with earlier (Section 2.2). The range of thermal diffusivities of nonmetallic liquids and solids is relatively narrow, namely between 0.15×10^{-6} and 2×10^{-6} m^2/s, while the thermal diffusivity of gases is about 20×10^{-6} m^2/s due to their low density, the value for metals is 10×10^{-6}–100×10^{-6} m^2/s. Examples are shown in Table 2.3.

It is interesting to note that the thermal diffusivity of composite heat insulators containing gases are close to the conductivities of the continuous phase because the presence of gas reduces not only the thermal conductivity but also the density while the heat capacity related to the unit of mass will vary only insignificantly.

In some engineering tasks dealing with liquids, the ratio of kinematic viscosity to thermal diffusivity, the so-called **Prandtl number** (Eq. 2.18) will play an important role.

While in gases the Prandtl number is somewhat smaller than 1, it is around 5 or more in common nonmetallic materials. In highly viscous materials, the Prandtl number approaches infinity while it is meaningless in solids. Substantially, smaller than 1 is the Prandtl number in molten metals because of their low viscosity and high thermal conductivity. For example, in lead and mercury, it is about 0.02 and in molten steel, it is 0.06.

2.7 Transport of mass by diffusion in liquids and solids

A quantitative measure of diffusion is the **diffusion coefficient** D, defined by relationship (1.5).

The **two-component diffusion** plays an important role in gases and vapors where the portions of both components A and B are high. In this case, component A moves into an area of low concentration of A while component B moves in the opposite direction. In this process it is possible that the density as well as the center of gravity move as well. This circumstance makes, of course, the mathematical modeling of this process very complicated.

If the diffusion involves the motion of a small amount of component A in a solid medium B, then rarely would the particles A change places with particles B. More frequent is the **interstitial diffusion** when the particles A "weave" between essentially motionless structure of mass B. Then, it is possible to view this case as a **one-component** diffusion of component A through a steady medium B from positions with a high concentration of A to positions having a low concentration of A while the concentration changes of B are insignificant.

2.7.1 Estimate of diffusivity of small molecules in liquids

According to several theories, the diffusion coefficient $\mathcal{D}_{AB}$ of a component A in a liquid medium B is directly proportional to temperature T and indirectly proportional to the dynamic viscosity of the liquid medium B:

$$\mathcal{D}_{AB} \sim \frac{T}{\mu_B} \tag{2.43}$$

Further, the diffusion coefficient rises with the increase of the space between the particles of the liquid medium B and it drops with the increasing size of the diffusing particle A.

A more complex relation was proposed, e.g., by Wilke and Chang (1955)

$$\mathcal{D}_{AB} = 1.17.10^{-4} \frac{T\sqrt{\varphi_B M_B}}{\mu_B V_A^{0.6}} \tag{2.44}$$

where temperature T is in Kelvin degrees, V_A [m^3/kmol] is the molar volume of the material A at its boiling point, μ_B [Pa s] is the dynamic viscosity of the liquid B, M_B [kg/kmol] is its molar mass and φ_B is an association factor. This factor is always larger than one when the molecules of B form larger aggregates.

The diffusion coefficient of small molecules in low viscosity liquids is of the order of 10^{-9} m^2/s while in gases it is about a 1000 times higher. The Schmidt number of liquids is usually higher than 1000.

2.7.2 Diffusion in solid materials

Interstitial diffusion in **crystalline materials** is unlikely.

The small atoms such as hydrogen and helium will diffuse easily in **metals**, particularly in heavy metals that are composed of large atoms. It needs to be emphasized that the hydrogen will be in the atomic mode and not in the form of molecules.

The diffusivity in **nonmetallic materials** will vary widely, usually in conformity with the peculiarities of the solubility of the A component.

This is well demonstrated in the case of polymers: while the PVC and the PMME are impermeable for oxygen, the silicon polymers and polymers with substituted fluorine will transmit oxygen exceedingly well.

2.7.3 Diffusion through membranes (flat layers)

In a solid material, it is convenient to introduce the more easily measurable equivalent concentration of the surrounding. To this end, it is necessary to know the **solubility**. Whole molecules will diffuse through **nonmetallic** materials, both liquid as well as solid and, concentration proportional to the concentration in the adjacent phase.

$$C_{in} = \frac{C_{out}}{a} \tag{2.45}$$

where a is an equilibrium constant. Specifically, when the outer medium is a gas, the concentration is proportional to the partial pressure

$$C_{in} = \frac{p_{out}}{H} \tag{2.46}$$

and H is the **Henry's constant.**

Somewhat different is the case in metals when the multiatomic molecules will dissociate in individual atoms at the entry in the metal. Then the diffusing atoms will share their electrons with the metal and, consequently, they will behave like the metal. Specifically, biatomic gases will follow the Sieverts law

$$C_{in} = \frac{\sqrt{p_{out}}}{H_S} \tag{2.47}$$

Thus if the concentration inside the layer is expressed in terms of the concentration in the surrounding medium then the basic equation for nonmetals (Eq. 2.44) becomes

$$n_x = -\frac{\mathcal{D}_{AB}}{H}\frac{dp_{out}}{dx} \tag{2.48}$$

in which neither the diffusivity $\mathcal{D}_{AB}$ nor the solubility H is usually known. Instead, however, their ratio, **permeability** of the layer

$$\mathcal{P} \equiv \frac{\mathcal{D}_{AB}}{H} \tag{2.49}$$

is used frequently to characterize the material properties of membranes.

For specific membranes, it is frequently impossible to determine their thickness. Therefore it is more useful to introduce another quantity known as the **membrane transmissibility**

$$\mathcal{T} \equiv \frac{\mathcal{P}}{\Delta x} = \frac{n_x}{-\Delta p_{out}} \tag{2.50}$$

It holds, therefore, that

$$n_x = -\mathcal{T}\,\Delta p_{out} \tag{2.51}$$

Thus the transmissibility, being the coefficient of the transfer of the material through the layer, expresses the amount of the material passing a unit area of the membrane per unit of time when driven by a unitary partial pressure step.

2.7.4 Thermodiffusion

Because the temperature gradient will influence also the concentration gradient, the consequence of the transfer of heat will be some material flow in mixtures. This phenomenon is known as the Soret phenomenon or the thermodiffusion. However, this phenomenon is not very pronounced and it can be neglected in most engineering situation.

A similar case of a dynamic equilibrium can occur when the flow of the material is accompanied by the transfer of momentum. This phenomenon will, however, become significant only in very large systems.

If there were no Earth attraction the atmosphere would escape into the Universe. As a result of the interplay of the gravity and the diffusivity, a dynamic equilibrium is maintained in the atmosphere so that a mixture of nitrogen and oxygen is prevalent in the troposphere. However, at 50 km above the Earth's surface, the nitrogen overbalances the oxygen and at 100 km above the surface, the atmosphere is made mainly of the light hydrogen with a small amount of helium. The concentration is, however, very low; less than 1 g of hydrogen per 1 m^3 *of the air.*

2.8 Summary

Transport phenomena are one of the manifestations of the mass structure. That is why the viscosity, the thermal conductivity, and the diffusivity depend on the behavior of the molecular systems.

All transports in gases are caused by the collisions of molecules. Therefore they depend on the mean velocity of the random motion of molecules and on their mean free path. The kinematic viscosity, $\upsilon \equiv \mu/\rho$, the thermal diffusivity $a \equiv \lambda/(\rho\ c_P)$, and the diffusivity $\mathcal{D}$, all have the same dimension of m^2/s, and, therefore their numerical values, under a certain set of state variables, are close. To a certain degree, these values can be derived from the molecular statistics. In general, these values grow with growing temperature and they are almost independent of pressure.

The random molecular motion in liquids is hindered and the transports take place only by the mutual touching of the molecules. The capability to predict the values of the transport coefficients is very limited. Therefore they have to be determined experimentally if they are not available in a tabular form in the technical literature.

The viscosity has no meaning in solid materials so that the momentum transfer is due only through a one-time deformation (elastic or plastic); this area is the subject of the materials engineering. The basic test methods here are slow traction and shear.

The study of the relationship between the tension and the deformation is the subject of rheology. This discipline covers not only the flow behavior of both liquids and solids but also any transient states. The time and the temperature enter here as important variables.

The thermal conductivity of solid materials is close to that this material has in a liquid state.

The diffusivity of solid materials is low and, therefore the transport of mass occurs mainly in thin layers and in membranes.

A special case is the transport in high vacuum where the collisions of the molecules with the containing walls prevail over the mutual collisions of the molecules (Knudsen flow).

A shock wave can occur at higher speeds than the rate of the transmission of pressure, which is at supersonic speeds.

References

Moore, W.J., 1999. Physical Chemistry, fifth ed. Prentice-Hall. ISBN 000209018X, 9780002090186.

Wilke, C.R., Chang, P., 1955. Correlation of diffusion coefficients in dilute solutions. AICHE J. 1, 264.

CHAPTER THREE

Transport phenomena at the interface

Contents

3.1 The working of the forces at the interface

Apart from the forces that hold together individual molecules (such as the chemical and ionic bonds), there are also substantially weaker intermolecular forces that can maintain the distances between molecules. They are known as the van der Waals forces named after the scientist who for the first time specified them and who, while making use of them, setup the model state equation of a real gas. One can learn more about these forces in physicochemical textbooks that deal with molecular kinetics (Moore, 1999).

The fact that van der Waals forces are both attractive and repellent helps us to explain a variety of properties of both solid and liquid materials. The origin of these forces lies in the intermolecular action of permanent or inductive electric dipoles. They become particularly significant at the interface.

Transport and Surface Phenomena
https://doi.org/10.1016/B978-0-12-818994-8.00003-8

3.1.1 The boundary between liquid phases

The resulting attraction between molecules in liquids is substantially higher than that in gases. Equally, the attraction between polar molecules is higher than that between nonpolar molecules. At the boundary of two liquid phases, there are always several rows of molecules, each of them affected by a neighboring row. Those molecules that have higher mutual attraction are drawn into the body of the liquid. As a result, the liquid in this region is more compressed and the layer contains a certain **interfacial energy** $\boldsymbol{W}$,

$$W_{interace} = \sigma S \tag{3.1}$$

that is directly proportional to the interfacial area S. The interfacial coefficient σ having the dimension N/m depends also on the neighboring layer. In the case of two liquids having limited miscibility or being almost immiscible, the coefficient σ depends on the nature of both liquids and also somewhat on the temperature. It is called **interfacial tension**. If the neighboring phase is air saturated with the vapor of the liquid then the coefficient σ takes the name of **surface tension**. Some examples of the surface and interface tension are presented in Table 3.1.

Table. 3.1 Surface and interfacial tension σ.

Surface tension at ambient temperature	N/m
Water	0.073
Mercury	0.474
Nonpolar organic liquids (hydrocarbons)	0.02–0.03
Organic liquids forming hydrogen bonds (e.g., alcohols)	0.03–0.07
Melted inorganic salts	0.05–0.25
Steel at 1600°C	1.8
Interfacial tension	
Water and nonpolar organic liquids	0.02
Partially miscible liquids	$\rightarrow 0$

Even though the interfacial forces are due to the existence of finite adjacent layers of molecules, the continuum theory permits to treat the boundary as being a geometrical area characterized by the surface or interfacial tension, respectively. The equilibrium will occur in those locations where the interfacial energy and the interfacial area S will reach a minimum. If the external forces (such as the gravity) are insignificant then the curved boundary will be spherical, subject to a compliance with the existing boundary conditions. Due to the tendency of the **curved boundary** to shrink, the pressure inside the spherical object will be higher than that on the outer side, irrespective of the position of the cohesive phase. If the pressure outside of the liquid sphere of a diameter d is p_0 then the pressure inside the sphere will be $p_0 + \Delta p$ where Δp is

$$\Delta p = \frac{4\sigma}{d} \tag{3.2}$$

For more complex interfaces the **Laplace relation** applies

$$\Delta p = \sigma\left(\frac{1}{R_1} + \frac{1}{R_2}\right) \tag{3.3}$$

where R_1 and R_2 are two mutually perpendicular radii of curvature of the boundary. This relation is frequently used to interpret the experimentally determined surface and interfacial tension.

3.1.2 Surface instability

The interfacial tension depends primarily on the composition of the two adjacent phases and on their temperature. The value of the interfacial tension can change in locations where transfer of heat and mass takes place. Then the material at the boundary can move from the positions of lower tension to those of higher tension. Such a motion can change the temperature and the composition at the boundary resulting in the increase of the interfacial tension difference. This phenomenon is known as the **Marangoni instability**. If one of the visualization techniques is used then one would observe separate, faintly pulsating regions. In the center of these regions, the tension reaches its maximum. This phenomenon can be used to enhance the nucleation of bubbles in boiling liquids.

3.1.3 Surface-active materials

In the mixture of two miscible liquids, there could be one that prefers to be closer to the boundary. Such a **surface-active agent (tenside)** for a water-oil boundary is soap, the molecule of which has at one end a hydrocarbon ending that is chemically bonded atoms, and a polar group on the other end, such as an anion of an organic acid with a metallic cation. The surface-active agent forms at the boundary a very thin layer in which the interfacial tension is significantly reduced. Such a layer might appear at concentration of the tenside as low as $1\,\mathrm{mg/m^2}$. Even a small amount of pollutants that cannot be analytically determined can act as a tenside that is very difficult to remove.

If the system is in equilibrium or if the changes in the shape of the surface are slow, then the concentration changes of the surface-active materials are not very pronounced; the experimentally determined interfacial tension should be sufficient to characterize the system.

If, however, the boundary is subject to tangential strain, as is the case during the flow, then the surface-active materials could be rolled over on the surface. This rolling motion is associated with a certain resistance that, qualitatively, could be seen as a surface elasticity, plasticity, or viscosity. In such a case, a higher concentration will result in the renewal of the surface. Unfortunately, the sole value of the surface tension will not be sufficient to interpret this phenomenon. To characterize the motion of drops, bubbles, and wavy surfaces, including their splitting and coalescence, just the knowledge of the surface tension is insufficient. To this day, no satisfactory generally applicable quantitative physical interpretation of the relevant influences was found.

In the past, the mariners used to pour oil into the stormy sea to change the rhythm of the waves and to facilitate the landing of the ship.

3.1.4 The boundary between a liquid and a solid material

Just like in the case of a boundary between two immiscible liquids, in the interface between a liquid and a solid material, attractive and repulsive forces compete with each other. These forces are generated in the molecular layer of the liquid adjacent to the layer of molecules in the solid phase. The significance of these forces can be assessed by the **wetting angle** ϑ that forms on the interface between liquid, solid, and gas. This angle forms according to the forces acting on the molecules of the liquid (Fig. 3.1).

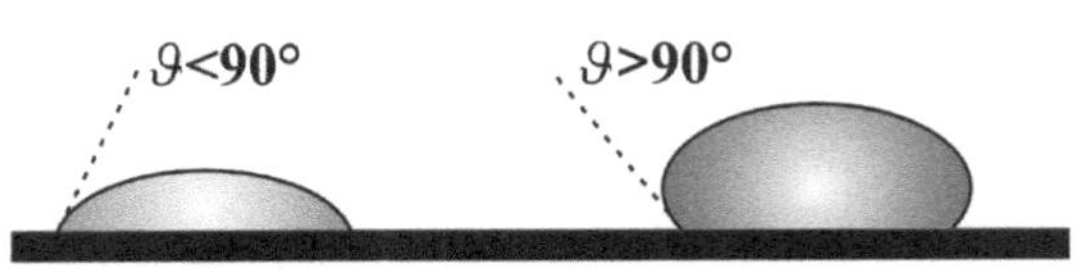

Fig. 3.1 A drop at the solid wall. Partially wettable; nonwettable.

A **wetting** occurs when the attractive forces in the liquid are weak and those on the surface of the solid are strong. In this case, the wetting angle is $\vartheta < 90$ degree. The extreme case when the wetting angle $\vartheta = 0$ degree is known as the **perfect wetting**. In the opposite case, when the molecules of the **nonwettable** liquid phase hold firmly together and the attraction by the solid phase is weak, then the wetting angle is $\vartheta > 90$ degree. In the extreme case, when the wetting angle is $\vartheta = 180$ degree. A wetting angle will form also at the contact of two immiscible liquids with a solid surface or with a gaseous phase. Also in these two cases, the surface acting agents, be they in the liquids or on the surface of the solid phase, will be determining.

As an example, let us consider the case of water in a glass. Clean glass is completely wettable by water. If, however, the surface of the glass is smeared by a non-polar material (e.g., a grease), then the glass surface becomes nonwettable and the drop of water will just stay on the glass surface. If, however, the water contains a surface acting agent (detergent), then it will freely spread on the glass surface. After some time, the grease with the surface acting agent will move to the boundary between water and air. Washing makes use of this phenomenon.

In the context of other transport phenomena, the wetting angle manifests itself particularly in the boundary conditions of the static phase boundary. In the case of faster processes (such as the sliding of a drop on a sloping surface), the equalization of the surface forces among the three phases is so slow that the wetting angles up and down the drop might differ slightly (Fig. 3.2).

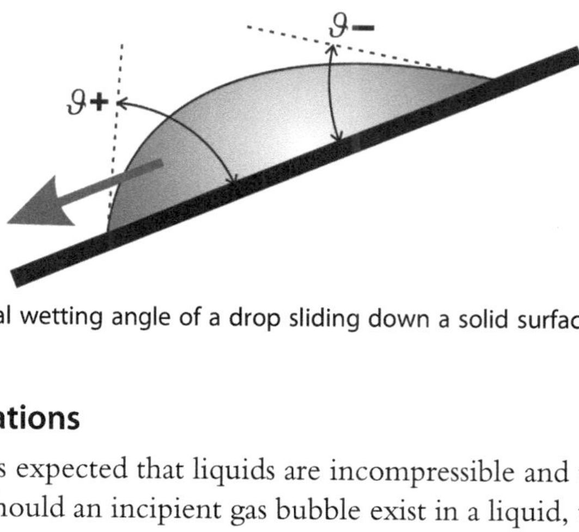

Fig. 3.2 Unequal wetting angle of a drop sliding down a solid surface.

3.1.5 Cavitations

Normally, it is expected that liquids are incompressible and therefore non-extendable. Should an incipient gas bubble exist in a liquid, then the liquid will evaporate into the bubble if the pressure in the bubble becomes smaller than the relevant vapor tension. This will result in the growth of the bubble while the positive pressure in the bubble will resist outside forces. Such a sudden pressure drop occurs in a flowing liquid behind a bend or an obstacle. In such positions, vacuum **cavities** will form if the cohesion of the liquid could be overcome and if enough energy is supplied to form an interface. These cavities will be gradually filled with low pressure vapor. While these cavities pulsate, they release small bubbles that escape into the main liquid flow where they disappear. This process might be accompanied by shocks and sound effects. In engineering practice, the **cavitation** is undesirable because it worsens the smooth running and the effectiveness of pumps, mixers, and turbines and increases the mechanical straining of the surrounding walls as well as their corrosion.

3.1.6 Other surface phenomena

There are other processes of rearranging materials that take place at the interface. They are, among others, adsorption on solid surfaces, **chemical reactions** in interfaces, and other **electrical, magnetic, and electrochemical phenomena**. The treatment of these subjects falls beyond the scope of this textbook. Interested parties are advised to consult the relevant literature.

3.2 Mechanical equilibrium between phases

3.2.1 Mechanical equilibrium at the interface of two fluids

The **normal tension** (i.e., **pressure**) has been already mentioned earlier. At an even boundary, it should be identical in both phases. However, the Laplace relation (3.3) will apply at curved boundaries.

To deal with **velocities**, it is advantageous to introduce a moving coordinate system such that the coordinate y, oriented in the normal direction, is equal to zero at the interface and the normal velocity u_y, has a zero value. The x coordinate coincides with the direction of the tangential motion. The coordinate z is indifferent. It follows from the mechanical equilibrium that there is an identical value of the **tensile stress** τ_{xy} at the boundary (i.e., $y=0$) between the two phases A and B

$$\left[\tau_{xy}\right]_A = \left[\tau_{xy}\right]_B \tag{3.4}$$

In this case a uniform coordinate system applies. However, from the point of view of the liquid boundary, the law of action and reaction applies so the directions of the stresses in the liquids are opposite.

The speed at any given time in the vicinity of any given point is a function $u_x(y)$ of the normal direction.

In the case of a **tangential motion of a liquid**, it is customary to assume that the **velocities** $u_x(0)$ are identical on both sides. It follows that the local velocity gradient

$$\mu_A \left[\frac{du_x}{dy}\right]_A = \mu_B \left[\frac{du_x}{dy}\right]_B \tag{3.5}$$

As long as $\mu_A \gg \mu_B$, as is the case when A is a liquid and B is a gas or A is a solid and B is a fluid, then, as a rule, the right-hand side of the above relationship can be neglected so that close to the boundary the velocity $[u_x]_A$ will be constant.

The intensity of energy dissipation, that is, the conversion of the mechanical energy into heat (the units being W/m^3), will then be

$$\epsilon = \mu \left(\frac{du_x}{dy} \right)^2 \tag{3.6}$$

It follows that the dissipation will be more significant in the fluid having a low viscosity.

It might be necessary to consider the dimensions of the boundary if dealing with systems having **very small dimensions**. It might be debatable where to place the hypothetical geometrical area separating the phases; should it be placed in the dips or on the spikes of a rough surface. In the case of dispersions, the distribution of the disperse phase in the liquid could play an important role. There is a tendency of the disperse particles to organize themselves in regular structures while, at the same time, a layer of a clear liquid forms at the boundary. This phenomenon manifests itself as an **apparent slip** at the wall.

3.2.2 Friction between two solid surfaces

Friction between two solid surfaces occurs when each of the two bodies has a different tangential velocity $u_x(0)$. Although the continuum mechanics assumes that the two interfaces are smooth geometrical planes, the reality is different. In reality, the two surfaces are **rough** even though they might, from a macroscopic point view, appear smooth. If a **normal pressing force** F_N is applied on the boundary then the spikes are partially deformed and pressed into the depressions. Regardless of whether the roughness is macroscopic or nanometric molecular in size, there will always be a limiting static tangential force that has to be overcome in order to bring the two solid bodies out of a stationary state. Without such a force, only a minor elastic deformation occurs. This phenomenon is known as **static friction** that, for two typical surfaces, can be characterized by a **coefficient of a drag static friction**

$$\mu_S = \frac{F_{T0}}{F_N} \tag{3.7}$$

It is remarkable that the limiting static tangential force is almost independent of the size and shape of the two planes or curved surfaces.

If the limiting static tangential force is exceeded, then the two bodies will slightly separate. This separation will result in the reduction of the spikes that hinder the mutual tangential motion. The consequence will be an accelerated tangential motion of the two bodies. Only a minor force F_{TK} is needed to maintain this motion after its initial start. This force is sufficient to maintain the two rough surfaces at a satisfactory separation so that it steadily overcomes the resistance due to the deformation of the roughness. This tangential force is also proportional to the normal pressing force that can be characterized by a dimensionless **coefficient of drag kinetic friction**

$$\mu_K = \frac{F_{TK}}{F_N} \tag{3.8}$$

Also this coefficient $\mu_K < \mu_S$ is almost independent of the size of the two contacting surfaces and, at low velocities, is independent of this velocity.

At this motion, a portion of the energy is dissipated in heat. This dissipation (units W) takes place in a very thin, almost negligible region of the contact area and it is determined by the product of the acting force and the relative shear velocity

$$W = F_{TK}|u_A - u_B| \tag{3.9}$$

Sometimes it is useful to relate this dissipation to a unit area of the two contacting bodies.

During kinetic friction, the tangential force is consumed for the deformation and jumps over the spikes of the two areas in contact. The situation is quite different in the case of **rolling friction**. In this case, the energy is used only for moving the spikes on the surface of the rolling body into the depressions on the underlying pad, on which the rolling body rolls. Possible deformation of spikes on both surfaces might also occur. It needs to be emphasized that the force needed to roll one surface over another is significantly smaller than the force needed for tangential friction.

In some cases the roughness can be **oriented**. This means that the friction coefficient is a vector that has a different value in different directions.

This phenomenon can be utilized in certain situations, such as the application of the so-called up-hill (seal-like) belts on skis or placing milled scales on the slip area. Other examples are flexible rings placed around screws to prevent their loosening.

The **reduction of the kinetic friction** is desirable in the bearings of rotary engines. This is achieved by choosing materials having very smooth surfaces. Another positive result is achieved by the combination of a very hard material with a soft one.

One such example in engineering applications is the combination of a hard steel with a soft bearing metal such as bronze. Also a plastic material could be used for bearings that are not highly strained. If the plastic is permanently damaged then all that is needed is the exchange the plastic insert during the next maintenance cycle. If both metallic surfaces of the bearing are hard then the bearing could be jammed and irreparably damaged.

A substantial reduction in the friction coefficient can be achieved by the use of **lubricants**. These are usually highly viscous liquids or low molecular plastics. The lubricant fills the depressions in the surface and prevents the spikes from snap. By a proper choice of the loading slope, the lubricant is pressed between the sliding surfaces thus helping to maintain the spacing between them. In the case of the **rolling friction**, pressure is generated in the lubricant in front of the rolling component. In the extreme case, the lubricant or even some gas is forced into the bearing by a pump. Then the pressure liquid exerts the normal force F_N. In this case, to talk about friction between solid materials is no longer appropriate (drones, hovercraft).

An opposite case occurs when slipping between solid surfaces is not desirable and an **increase in friction** is desired.

A typical case is the contact of the tread lug of a tire with the surface of the road. It is obvious that the friction will be higher the deeper the grooves into which the ridges fit will be. It is, therefore, required for the tires to have sufficient depth of the grooves. Further, the tires must have proper flexibility which is achieved by choosing a rubber mixture with a lower modulus of flexibility for the winter season while in summer a harder mixture is preferred.

Sufficient friction is required from the so-called **friction materials**. They are expected to smoothly transfer the tangential forces as is the case when braking the vehicle or when pressing initially the clutch. A secondary task is to completely stop the mutual sliding of two solid materials. It is necessary to take into account that the kinetic friction is smaller than the static one. This fact may bring about an instability at which, in a certain range of forces, the sudden changes from a static to a kinetic state might occur. This case might be manifested by sound effects like rattling and creaking.

When braking a motor vehicle it is desirable to maintain the static friction on the surface of the roadway while maximizing the tangential force of the kinetic friction in the brakes. This is achieved by a judicial choice of friction materials and also by using the so-called anti-block system (ABS) that automatically insures that braking force is evenly applied to all braking components.

Friction surfaces acting against metallic surfaces are usually rough particles embedded in a hardened resin. Also included might be metallic particles, such as copper wires, the role of which is to lead the friction heat away from the brake surface. The micro- and nanoparticles released during braking from the friction materials might be an environmental problem. For this reason asbestos is no longer permitted.

3.3 Heat equilibrium between phases

The measure of the heat equilibrium is temperature. The temperatures close to a boundary between two phases A and B are

$$T_A = T_B \tag{3.10}$$

Also the heat flow across the boundary at equilibrium is

$$q_A = q_B \tag{3.11}$$

Should the heat flow be simply due to conduction, then

$$\lambda_A \frac{dT_A}{dn_A} = -\lambda_B \frac{dT_B}{dn_B} \tag{3.12}$$

where n is the normal perpendicular to the interface and pointing toward the given phase.

The heat flows q_A and q_B will differ only when the heat is consumed in the interface. This case occurs when the heat is consumed or created by friction, the change of the state of aggregation, or by a chemical reaction running in the interface. In this case, the change in the heat flow is related to a unit area. Should the interface move, it will move only very slowly. If a phase change occurs in the volume then the heat of aggregation or the reaction heat is related to unit volume.

3.4 Equilibrium concentrations between phases

When the phases are in equilibrium then the so-called Gibbs rule applies

$$\begin{aligned}&\text{The number of the degrees of freedom } (F)\\ &\quad = \text{the number of components } (C) - \text{the number of phases } (P) + 2\end{aligned} \tag{3.13}$$

The **phase equilibrium** in a system having **one component** ($C=1$) and **two phases** ($P=2$), out of which one is vapor, will have 1 degree of freedom ($F=1$). This means that from a known temperature the pressure $p_A^0(T)$, known as the **vapor tension** at a given temperature, can be determined. Inversely, should the pressure be known then the dependence of the **boiling temperature** on pressure will be

$$T_{V,A}(p) \tag{3.14}$$

In the case of two **condensed phases** (i.e., there is no gas phase present) the system is, under normal conditions, almost incompressible so the pressure does not represent one of the degrees of freedom, then the system will be in equilibrium at only one temperature and that is the **melting point** $T_{T,A}$.

The expression of a system composition in multicomponent multiphase systems is more problematic since there are numerous options derived from specific customs. One such option could be the way a mixture is prepared or the analytical determination of the composition. Yet another option is based on the usefulness for subsequent calculations.

The composition, being an intensive quantity, could be expressed as the ratio of the amount of the tracked component to

– total amount of the mixture, or
– the amount of another component.

The amounts can be expressed either

– as the mass (mol),
– as the weight (kg),
– as the volume (m^3),
– as partial pressure, or
– others.

The ratios of quantities having the same dimensions can be expressed in fractions, for example, percentage or ppm (=parts per million).

Somewhat confusing is the unit ppb (=parts per billion) since in US English the billion is the European milliard.

While performing balance calculations it is often advantageous to reduce the compositions of individual components to one basis, such as the mass fractions or the mole fractions. To do this, one needs to know the respective definitions of these variables.

The composition may be related to some conventional ratio. Here are some examples:

the relative humidity is the ratio of the actual mass fraction of the water vapor to that of a saturated air at the same temperature and pressure, another obscure quantity is the volume percentage of alcohol in drinks. In reality, this is the volume of pure alcohol at 20°C present in 100 volumes of the drink, and many others.

To perform these calculations one needs to know more physicochemical data.

The **two-component equilibrium** in a system with two phases out of which one is a **vapor** will have 2 degrees of freedom ($F=2$). One of them could be the composition of the liquid phase (the molar fraction of the component A in the liquid, x_A) and the temperature T. An approximate relationship that binds these variables is the **Raoult law**

$$y_A \equiv \frac{p_A}{p} \approx p_A^0(T)\, x_A \tag{3.15}$$

where p_A is the partial pressure of the component A in the vapor phase and p is the total pressure. When adding the equivalent relationship for the component B and when using the relevant definitions, one obtains

$$\begin{aligned} p_A + p_B &= p \\ x_A + x_B &= 1 \\ y_A + y_B &= 1 \end{aligned} \tag{3.16}$$

The remaining unknown quantities can then be determined by performing

$$y_A \equiv \frac{p_A}{p} = \frac{\alpha x_A}{1 - x_A + \alpha x_A} \tag{3.17}$$

where α is the **relative volatility**.

$$\alpha \equiv \frac{p_A^0(T)}{p_B^0(T)} \tag{3.18}$$

Interestingly, the relative volatility of a number of mixtures of similar materials depends only weakly on temperature so it could be considered a constant for a given pair of substances while performing engineering calculations.

The **real mixtures** do not usually follow the Raoult law. The physical chemistry bypasses this problem by defining new variables, specifically the activity and the fugacity. A chemical engineer, who is usually concerned with problems of distillation, will rather use experimentally determined relation of x_A and y_A that might have the form of a generalized relationship on temperature and pressure or that can be obtained from tables and graphs.

In the case of the equilibrium between a **gas** and a **liquid** the relative volatility is very high. Then the main component of the gas is A ($y_A \to 1$)

and the main component of the liquid is B (thus x_A is very small). The relationship (3.17) is then reduced to a linear relation characterizing the **solubility of the gas** in the liquid that is the **Henry law**

$$x_A = \frac{p_A}{H(T)} \tag{3.19}$$

The equilibrium between a **gas and a solid phase** and some of its peculiarities were already treated in Section 2.7 in conjunction with the diffusion.

Although the **two-component** equilibrium in a system with **two condensed phases** has 2 degrees of freedom ($F=2$), the influence of the pressure can be neglected. This is the case of the dissolving of crystals or the reverse process of crystallization. In this case, the key information is the knowledge of the **solubility** in the chosen solvent. The **relative percentage concentrations** of saturated solutions (usually presented as the mass of the solid phase in grams dissolved in 100 g of the solvent) are published in tabular form. For practical reasons, the tabulated data contain usually the solubilities at ambient temperature and at the boiling point.

The so-called **solubility product** K_S is usually used for poorly soluble ionic compounds. For a substance A_aB_b, while dissolving, splits into

$$A_aB_b \rightarrow aA + bB \tag{3.20}$$

the solubility product K_S is defined as

$$K_S \equiv [A]^a[B]^b \tag{3.21}$$

The square brackets in this relation mean the numerical value of the concentration of a given substance (e.g., cation or anion) in mol/dm^3.

3.5 Summary

There is equilibrium at the interface between two distinct phases.

The mechanical equilibrium is expressed as the numerical identity of the tensile stresses. They will have, however, opposite signs.

Apart from the normal forces (pressure) between two fluids, an interfacial force also comes in the picture if the interface is curved. This is the interfacial force. In the special case of a liquid in contact with the air, the tension is known as the surface tension.

Heat equilibrium between two phases is characterized by the identity of temperatures in corresponding points of contact.

The concentration equilibrium is determined solely by the thermodynamic equilibrium even if transport processes take place. Formulas exist that allow the determination of the relationships between individual concentrations from the phase or chemical equilibriums. Simple formulas are usually available for the ideal mixture along with the appropriate corrections. Further, tabulated data exist for the nonideal mixtures.

Reference

Moore, W.J., 1999. Physical Chemistry, fifth ed. Prentice-Hall. ISBN 000209018X, 9780002090186.

CHAPTER FOUR

Transport of energy by radiation

Contents

4.1 Types of radiation

The surface of any body may emit or absorb energy in the form of an electromagnetic radiation at various wavelengths. The character and the amount of this radiation depend on the nature of the surface (i.e., its chemical composition) and the mode by which the radiation is initiated. Although there is a continuous progression of waves of different wavelengths it is convenient to classify this radiation into several classes (see Table 4.1).

Table 4.1 Classes of radiation.

Wavelength	Class
<0.1 nm	Gamma (γ)
<10 nm	X-rays (XR)
<400 nm	Ultraviolet (UV)
400–800 nm	Visible
0.8–1000 μm	Infrared (IR)
8–12 μm	Heat
<10 mm	Microwaves
<10 km	Radio

Specifically, we shall concentrate on **heat radiation** that occurs between two solid surfaces having significantly different temperatures.

Transport and Surface Phenomena
https://doi.org/10.1016/B978-0-12-818994-8.00004-X

4.2 Heat radiation and its absorption

Heat radiation becomes significant only at high temperatures when it can be sensed both as heat as well as light. At low temperature its role is usually negligible. Unlike in the case of transport phenomena due to the motion of molecules, the transport by radiation is almost instantaneous moving the energy at the speed of light, including in a vacuum. While this dictum applies to all kinds of radiation we shall focus exclusively on infrared radiation that facilitates transport of heat. Although the transport of heat by radiation is not governed by linear mathematical models like all the other transports, it is convenient to deal with radiation in this context since, to some extent, it competes with the conduction of heat.

The amount of **heat emitted** as well as **absorbed by radiation** is governed by the **Stefan-Boltzmann law** which states that the intensity of heat flow by radiation is

$$q = \sigma \epsilon T^4 \tag{4.1}$$

where the Stefan-Boltzmann constant is

$$\sigma = 5.67 \times 10^{-8}\,\mathrm{W}/\left(\mathrm{m}^2\mathrm{K}^{-4}\right) \tag{4.2}$$

T is the absolute temperature and ε is the degree of emissivity.

The degree of emissivity ε of a perfect black body is equal to 1 while the emissivity for a mirror or a glossy surface is 0.02.

Thus, the transport of heat by radiation in a vacuum between two parallel **black surfaces** having temperatures T_1 and T_2 is

$$q = \sigma\left(T_1^4 - T_2^4\right) \tag{4.3}$$

Radiated and absorbed heat depend on the size of the projection of the mutually facing surfaces. In such a case, appropriate formulas and diagrams can be found for individual geometric setups in respective handbooks. They fall outside the scope of this monograph.

In nonideal cases, the surfaces posses a various degree of grayness, which means that they are not ideally black or completely glossy. They reflect part of the impacting radiation and partly absorb it. The extent of the reflected radiation depends on the wavelength of the impacting radiation.

A remarkable case is the reflection from the surface of freshly fallen snow, which reflects about 90% of the impacting light at a wavelength of about 0.5 μm, that is, in the range of the visible light. However, for the heat radiation at a wavelength of about

10 μm, the snow surface behaves as an almost ideally black body having $\varepsilon = 0.97$*. A similar case applies to paper* ($\varepsilon = 0.87$) *and glass* ($\varepsilon = 0.95$)*.*

4.3 Transmission of heat by radiation through a mass medium

Should the space between the surfaces, involved in exchanging heat by radiation, contain **transcalescent** mass then the molecules of this medium may participate in the process. They may allow the heat to pass thus returning the heat by radiating it back in all directions. Although some materials with high effectiveness may allow the heat radiation to pass, the heat flow will be always impeded by them (e.g., water or glass). High-intensity heat rays originating from a source of higher temperature (low wavelength) will penetrate the medium better than the reflected rays from a surface that has a lower temperature. When the surface having lower temperature is the surface of the Earth then this phenomenon that occurs is known as the **greenhouse effect**.

While some gases (particularly simple gases having molecules with only one or two atoms, such as N_2, O_2, Ar) are essentially inert toward heat radiation, other gases (such as H_2O, CO_2, O_3, NO_x, CH_4, and others) can absorb, as well as emit, a considerable amount of heat. This circumstance is of considerable importance in engineering practice because the heat from the combustion gases (containing mainly H_2O and CO_2) can be transmitted to the wall of a furnace or of a boiler not only by conduction from the flowing gases but also by radiation.

Gases with multi-atomic molecules that occur naturally in the Earth's atmosphere bring about the so-called greenhouse effect. This phenomenon occurs because the temperature of the Earth's surface that is determined mainly by the influx of heat from a relatively small, and only a part of the day, visible area of the Sun (temperature of which is 5800 K and area about 1/100,000 of the sky), is balanced by the radiation (both during the night as well as during the day) into the whole universe, the temperature of which is 0 K. Thus, the surface of the Moon or Mars is significantly cooler than the surface of Venus that is shrouded by a massive layer of CO_2*. The extent of the influence of the greenhouse gases is determined by the kind of the molecule and its concentration. The so-called global warming potential (GWP) indicates the concentration of* CO_2 *which would have the same potential as the unit concentration of the medium considered, for example, the GWP for* CH_4 *is 34 and for* N_2O *298. Currently,* H_2O *contributes to about 36%–60% of the greenhouse effect,* CO_2 *about 9%–26%,* CH_4 *about 4%–9%, and* O_3 *about 3%–7%. Other gases in trace amounts may also influence the effect. Also, the variation in the atmospheric* *humidity, both in the form of vapor and condensed clouds, may play a role. The recent decades have witnessed an intense debate on extent to which the industrial activities releasing CO_2 are responsible for the increase of the greenhouse gases in the atmosphere.*

4.4 Radiation in chemical analysis

Selective **absorption** or **emission** of radiation of different wavelengths can be used to develop **spectrometric methods** of chemical analysis to determine the presence of specific molecules. From the intensity of their occurrence one can determine their quantitative representation.

4.5 Summary

The transmission of heat by radiation depends pronouncedly on temperature, specifically on the fourth power of the absolute temperature. Both warm as well as cold bodies radiate heat. The amount of heat radiated depends on the difference of the fourth powers of absolute temperatures.

The amount of heat transmitted by radiation is proportional to degree of emissivity ε (blackness) which depends on the nature of the radiating surface. The emissivity for an ideally black body is equal to one while it approaches zero ($\varepsilon = 0.02$) for glossy surfaces.

Any surface will appear blacker toward the heat radiation than toward the visible light. Media appear blacker toward the heat radiation than toward the visible light. Similarly, the transmission of heat is harder than the transmission of light through the media (with the exception of gases consisting of molecules having one or two atoms). All other gases and liquids (including glass) have higher resistance to the transmission of heat, particularly at lower temperatures (higher wavelengths). This phenomenon is known as the greenhouse effect.

Radiation of different wavelengths behaves differently and this effect can be utilized in the development of analytical methods. The differences in light transmissivity can be used to develop absorption spectrometry. The emission spectrometry is based on the variation in radiation from surfaces.

CHAPTER FIVE

Experimental methods for the study of transport phenomena

Contents

Originally, the methods used to determine the transport properties were based on the measurement of some mechanical signals, the changes of which would have been observed **visually**. However, for convenience, the tendency was to convert the mechanical signals to either an electric or a pneumatic output that would enable a **remote control** of the property, for example, by a voltmeter. Should the measurement of the required property be replaced by the measurement of a better accessible property then a **theoretical relationship** between these two properties must exist. Alternatively, the measuring device could be **calibrated** by a known **standard**.

At the end of the 18th century, these standards were determined by an international agreement in relation to some independently existing physical realities, of course with an accuracy possible at that time: the unit of length, 1 m, was set in relation to the circumference of the Earth, the unit mass, 1 kg, was related to the density of water, and the unit of time, 1 s, to the length of the day. The physical models of these standards are deposited in the International Bureau of Weights and Measures in Sevres, France. Currently, new standards derived from naturally existing quantities with an accuracy of 10 decimal places are internationally recognized. They are, for example, the wavelength of the radiation of the atom of cesium, the speed of light, the Boltzmann constant, the Plank constant, the triple point temperature, and others.

Transport and Surface Phenomena
https://doi.org/10.1016/B978-0-12-818994-8.00005-1

5.1 Mechanical quantities

5.1.1 Force

The unit of force in the SI system is Newton, $1\,N = 1\,kg\,m/s^2$, which is a force that imparts to the mass of 1 kg an acceleration of $1\,m/s^2$.

The older unit was based on Earth's gravity. One kilogram force was equal to the weight of 1 kg mass. Later, this unit was called kilopond (kp). With the use of the standard gravity, $1\,kp = 9.80665\,N$.

The weight of a known mass is still being used for the calibration of dynamometers. **Scales** can be used for the measurement of force provided that it could be, with the help of levers and pulleys, redirected vertically. The measured force is then manifested as a load or unload on the pan of the scale.

The standard **dynamometry** used visibly deformable elements that followed Hooke's law. These elements were usually made in the form of a coil from a hardened, spiral-quality steel or from phosphorous or beryllium bronze. For a torsion force, the element takes the shape of a helix. For the smallest, very sensitive measurement, a torsion wire would provide a satisfactory sensing device.

Modern dynamometers use elements that undergo only a very small deformation that they convert to electrical signals. Usually, these devices respond very promptly to the applied force. **Tensometers** are usually cemented to the surface of the deformed element. They contain in their body a fine electrical conductor that changes its electrical resistance when exposed to even an infinitesimal deformation. The use of **piezzo crystals** is based on their property to change the measurable electric charge on their surface when the inner electric charge moves due to the imposition of a compression.

A slight deformation of a wall forming a gap with a sensor will bring about a change in the **electric capacitance** of the condenser formed by such gap.

5.1.2 Pressure

Pressure is the force acting on a unit area. Its unit is Pascal, $1\,Pa = 1\,N/m^2$. The pressure on a spot on the wall can be determined by measuring the force acting on this spot. To this end, the spot can be provided either by a piston or by a flexible element. To measure the force on this spot, a spring, a tensometer, or a piezoelectric sensor can be used. Currently, miniature probes are available which can be immersed even in liquid media.

Alternatively, the pressure can be also measured by branching the fluid in an external manometer that contains a flexible element, the deformation of which can be monitored. In essence, this device determines the difference between the pressure in the measured medium and the pressure in the surrounding space. The simplest manometer consists of spiral tube that straightens up under pressure and this movement is visualized by the motion of the needle of the manometer. Another alternative is a spring dynamometer controlled by a piston.

The force induced by an external pressure can be compensated for by raising the level of the measuring liquid. A frequent task is to measure the pressure difference between two spots in the liquid. The simplest way to go about it is to use a transparent tube in which the liquid separates the gas, taken from two different spots. As a result of the pressure difference, the liquid will rise to two different heights. Should the pressure difference be measured in a liquid then the separation can be achieved by gas or another immiscible liquid.

A method to measure atmospheric pressure was known from the times of Torricelli (1608–1647). Torricelli used a **barometer** consisting of a glass tube filled with mercury that was tightly closed from the top. As a result of this setup, vacuum forms above the level of the mercury so the pressure on the measured site will be equal to the height of the mercury column.

At an ambient temperature, the saturated vapor pressure of mercury is about 0.2 Pa, which corresponds to about 0.001 mm of hydrostatic pressure of mercury. Such a pressure is immeasurable. At higher temperatures, the mercury barometer is not being used as the vapor pressure of mercury might become significant.

A standard atmospheric pressure was chosen by an international agreement which corresponds to a column of 760 mm mercury (sometimes known as 760 Torr). In SI units, this is equal to 101,325 Pa. In the past, this pressure was also known as the physical atmosphere.

Another unit of pressure used frequently in the past was "the technical atmosphere" which is equal to 1 kg of weight acting on one square centimeter. In SI units, it is equal to 98,100 Pa. Yet another unit used mainly in the English speaking countries is 1 psi (pound square inch), which is equal to 6895 Pa, defined as the pressure resulting from force of one pound on an area of a square inch.

To measure pressures close to the atmospheric pressure, another unit "bar," by international agreement, is used; 1 bar is equal to 100,000 Pa.

Among the mechanical barometers the **aneroid barometer** measures the absolute pressure by evaluating the deformation of a metal membrane serving as a lid of a hermetically closed container.

5.1.3 Speed

The most frequently used method for determining the **speed of solid bodies** is the measurement of distance over a period of time. The instantaneous speed can be obtained by a continuous measurement of the frequency of the revolutions of the body rolling on a solid pad (car wheel). The rotational speed can be determined by coupling the rotating body with a dynamo. The rotational speed then can be determined from the output of the dynamo, which is dependent on the speed. Alternatively, digital determination can be made by counting the pulses of a perforated wheel.

The **mean speed of a flowing fluid** is usually determined by the measurement of the volume passed through a given cross section. Either a volumetric device or a calibrated sensor of the dynamic pressure can serve as a flow meter. However, the most common method is based on the measurement of the **pressure drop** occurring on a device generating significant hydraulic resistance, such as an orifice, a venturi tube, or a rotameter. Induction sensors can be used to convert the reading to an electric signal.

The basic factor for the selection of the correct device to measure the **local flow velocity** is the size of the space in which the data should be obtained. **Anemometers** consisting of a propeller and a scanner of the rotational speed are used to measure the wind speed and the rate of ocean currents. These could be relatively large devices. Somewhat finer **hot wire anemometer (HWA)** consists of a glowing wire, the temperature of which, and therefore also the electrical resistance, depends on the speed and the temperature of the circumfluous fluid. A relatively simple method makes the use of the **Pitot tube** that consists of a tube, the opening of which is oriented against the flow of the fluid. In agreement with the Bernoulli equation, the pressure inside the tube will rise with the second power of the flow velocity.

In transparent fluids, accurate measurement of local velocity could be made by tracking a tracer. There are two basic techniques that can be applied. The technique most frequently used is the **laser Doppler anemometry** (LDA). This method is based on crossing of two laser beams at a point where the velocity is to be measured. If a tracer particle passes through this point then the change in the frequency of the reflected beam will depend on the speed of the particle (Doppler principle). This method is so sensitive that it could be used to measure turbulent oscillations in the flowing fluid. It is generally assumed that there is no slippage of the fluid on the surface of the tracer.

Another tracer-based method is the **particle image velocimetry** (PIV). Here, laser beam (known as the "laser knife") is used to illuminate an area in the flowing liquid so that a multitude of particle positions can be monitored. If a sequence of such pictures is available then this sequence can be processed by a dedicated software to obtain the local velocity field.

5.1.4 Viscosity

To determine the viscosity, one needs to know the force acting on a surface due to the flowing liquid and also the velocity distribution on the surface. Usually, it is necessary to know, or to give, a dynamic quantity (such as the force) and to measure some kinematic variable (such as the speed). Inversely, it is possible to measure the dynamic effect of a given kinematic quantity. To back calculate the detailed values in the chosen surface, some kind of a balance needs to be applied. In simple cases of the so-called viscometric flows, this would lead to an ordinary differential equation. Should the flow pattern be more complex, a partial differential equation may result. Alternatively, it might be possible to start an experimental calibration of a specific measuring setup by a liquid having a known viscosity. This must be chosen so that it would be directly proportional to the dynamic quantity (given or measured) and indirectly proportional to the kinematic quantity (given or measured). Viscometric setups are dealt with in Chapters 7 and 12.

5.2 Thermal quantities

5.2.1 Temperature

The precise thermodynamic definition based on the knowledge of the Carnot cycle is not usable. Therefore the temperature is determined by the measurement of the changes in some of the quantities that vary with temperature. Today, a number of standard thermometric sensors for specific temperature ranges are available along with methods to convert the measured values to sufficiently accurate temperatures.

The standard **thermometer** makes the use of the **extensibility of liquids**. It is usually a flask connected to a glass capillary. Because the heat extensibility of liquids is substantially larger than that of glass, the level of the liquid in the capillary will move with the variation in the temperature. For years, mercury has been used as the measuring liquid. However, mercury cannot be used for high temperatures since its boiling point is 356°C. The advantage of using mercury is its electrical conductivity. Due to this circumstance, mercury could be used as an electrical switch when the temperature reaches a set point. Its disadvantage is its poisonous nature.

Another advantage of using mercury as the thermometric liquid is the ease with which the level of the mercury column indicates the highest temperature to which the thermometer was heated. Since the mercury column can be shaken down the thermometer can be reused for another measurement. For lower temperature, it is convenient to use dyed alcohol since the freezing point of mercury is −39°C.

The heat **extensibility of solids** is utilized for the construction of **bimetals**. These are made of two metal strips having different heat extensibility. On heating, the two strips will extend differently causing the bimetal to bend and the extent of the bending can be measured.

A **gas thermometer** was used at very low temperatures. The temperature is determined from the measured pressure and volume of a gas (usually hydrogen) by substituting these values in a suitable equation of state.

In the past, very high temperatures were determined by using the so-called **melting cones** (also known as pyrometric cones or Seger cones) made of a mixture of oxides to have various fusibilities. From their bending it was then possible to estimate the range of temperatures.

Currently, the modern methods are based on a change in electric properties with temperature.

An **optical pyrometer**, frequently used to measure very high temperatures, works on the principle of comparing the color of a measured surface with the color of a red-hot wire. A more advanced technique, the **infrared pyrometer**, measures directly the infrared radiation from a hot surface. It is even possible to measure a less intensive radiation from surfaces at ambient temperatures by means of a **radiation pyrometer**. Using the so-called **thermovision** the temperature distribution of a surface can be displayed in two-dimensional (2D) images.

A resistance thermometer equipped with a platinum wire provides a very accurate measurement. Less accurate is the **thermistor** exploiting pronounced changes in the electrical resistance of oxidous materials. It facilitates much more accurate and convenient measurement than that obtained by the determination of the **electrical resistance** or tension with the help of voltmeters, ampermeters, and resistance standards.

Another electrical phenomenon influenced by temperature is **thermoelectric** phenomenon that occurs in the boundary between two different metals. At such boundary, an electrical potential arises when exposed to different temperatures. This is because, at different temperatures the electrons in different metals have different mobility so that move from metal to the other. This phenomenon is applied in **thermocouple** sensors.

5.2.2 Heat capacity

The change in the heat energy, that is, the change in the **heat capacity** and the **latent heat** of the phase change or the **reaction heat** is measured using **calorimetry** where the heat transferred to a known object is balanced while the changes in temperature are measured. It is relatively easy to determine the amount of heat required for a phase change of a reference liquid provided that the measurement is done close to the particular temperature. Specifically, either the amount of liquid condensed from a saturated vapor or the amount of liquid generated by the melting of ice can be measured.

Heat capacities depend somewhat on temperature. The constants of several empirical formulas for the calculation of heat capacities are tabulated in the available literature (NIST database). They are given in SI units of J/(mol K); however, more frequently used units in engineering practice are kJ/(kg K). Before the wide acceptance of SI units, the units often used were kcal and BTU for water $c = 1\,\text{kcal}/(\text{kg K}) = 1\,\text{BTU}/(\text{lb}\ {}^\circ\text{F})$.

Thermal analysis tracks the increase in temperature of a specimen that is being slowly heated up at a steady rate. This information then can be used to determine the temperatures of phase changes and chemical reactions. In addition, this method also offers a way to estimate the heat orientation of chemical reactions.

5.2.3 Thermal conductivity

The determination of thermal conductivity requires the knowledge of the heat flow along with the knowledge of the temperature difference in a chosen geometrical setup. Usually, one of these quantities is given and the other is measured. The easiest way is the measurement of thermal conductivity of solid materials where a large temperature difference can be applied thus minimizing possible error in the determination of the heat flow. The chosen geometrical setup for the measurement is either set of concentric sphere or coaxial cylinders where the influence of sides could be eliminated.

Much more difficult is the determination of heat conductivities of liquids and gases due to the influence of natural convection. This problem can be somewhat alleviated by choosing suitable geometrical configuration. This could be two parallel plates oriented horizontally where the heated plate is up. The heat losses to the sides must be taken care of by making sure that the heat balance on the considered section of the plate was correct. The determination of the thermal conductivity of fluids is performed only in leading laboratories.

5.3 Summary

The physical properties are, as a rule, physical concepts that cannot be directly measured. What, however, can be observed are their quantitative as well qualitative manifestations.

The term "measurement" can be understood as a numerical comparison of a quantity (e.g., length) with a standard scale. This comparison will result in a quantity having a numerical value and a unit. Some properties cannot be measured directly. However, these can usually be determined by calculating the properties that are directly measurable.

In the past, the standard measurement that made use of visible manifestations of some related properties was most frequently the length (e.g., the length of a water column for the determination of the pressure, volume for the determination of temperature, or swings of a pendulum for the determination of time).

It is convenient to convert these properties to electric signals, which makes it possible to digitize simple quantities, possibly also of quantities not so simple.

CHAPTER SIX

Simple problems in heat transport

Contents

Transport phenomena play a significant role in a number of processes. As long as they take place in a simple geometric setup it is possible to derive relationships that govern them. In most cases, these are simple ordinary differential equations. Solving those results in correct conclusions. These will be dealt with in the following section where some heat conduction problems will be demonstrated. Solutions of selected flow problems will come next. These will, however, require the application of vector calculus.

The real-life problems take place, however, in a more complex geometric space and, in addition, often depend on time. The relevant mathematical balances in three-dimensional space and time that lead to a set of partial differential equations are derived in Chapter 16. In most cases, the advanced users no longer attempt to derive the relevant relationships ab initio. Instead, they proceed backwards by simplifying these relationships by leaving out the zero terms in the partial differential equations. In this way, they will proceed directly to the mathematical model of the problem at hand in the form of equations that are easier to solve.

Transport and Surface Phenomena
https://doi.org/10.1016/B978-0-12-818994-8.00006-3

6.1 One-dimensional steady conduction of heat

The principles of transport phenomena and the basic definitions will be first elucidated on a simple case of one-dimensional heat conduction case. Some simple problems will be outlined and their solution will be indicated. In most cases, it is possible to find a mathematically analogous case solvable by the same tools.

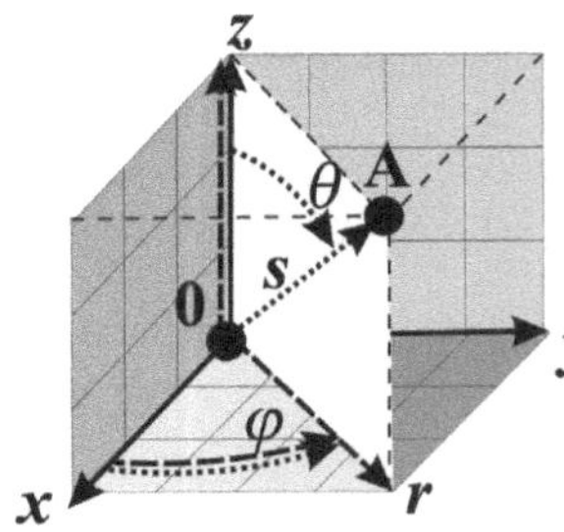

Fig. 6.1 Coordinates of the point A (The zero point is 0). (———— in print version) Cartesian *x, y, z*; (-------- in print version) cylindrical *r, z,* φ; (........... in print version) spherical *s,* φ, ϑ.

6.1.1 Coordinate system

The one-dimensional conduction realized in a **Cartesian coordinate system** x, y, z can occur in an arbitrary direction. However, it is a common practice to assign the x to the direction in which the heat is conducted. In the **cylindrical coordinate system** r, z, φ, and **spherical coordinate system** s, φ, ϑ, the conduction in the r or s direction (Fig. 6.1) is a specific case. In this specific case, the one-dimensional conduction in the z direction does not imply symmetry as is in the cylindrical system. If the conduction were directed along the φ or ϑ coordinates, then the motion in the r or s direction would occur automatically and the flow would not be one-dimensional flow.

6.1.2 Temperature distribution in a flat plate conducting heat

The equation of heat conduction

$$q_x = -k\frac{dT}{dx} \tag{6.1}$$

can be solved directly. Since the heat flux q does not depend on x in the case of a steady (no accumulation) unidirectional conduction through a flat plate, the solution of this differential equation will be a linear relationship

$$T(x) = -\frac{q_x}{k}x + C \tag{6.2}$$

where the integration constant will be determined from the boundary conditions. As a rule, the boundary condition for heat transfer problems has one of the two forms:

– Boundary condition of the **first kind** (Dirichlet's) when the value of the function (in this case the temperature) is known at a specific position, for example,

$$\text{for } x = x_i \text{ is } T = T_i \tag{6.3}$$

– Boundary condition of **the second kind** (Neumann's) when the derivation of the function (derivation of the temperature or the value of the heat flux) is known at one specific position, for example,

$$\text{for } x = x_0 \text{ is } \frac{dT}{dx} = -\frac{q_x}{k} \tag{6.4}$$

In either of these two conditions, the temperature distribution is given by the relationship

$$T(x) = T_0 - \frac{q_x}{k}(x - x_0) \tag{6.5}$$

in which one unknown quantity remains.

The knowledge of both $q(x)$ and $T(0)$ is needed for a complete description of the temperature profile. Alternatively, next to $T(0)$ another temperature $T(1)$ at another position must be known, possibly on the coordinate $x(1)$. It follows for this case that

$$T(x) = T_0 + \frac{x - x_0}{h}(T_1 - T_0) \tag{6.6}$$

The heat path h is here identical with the thickness of the slab

$$h \equiv x_1 - x_0 \tag{6.7}$$

The heat flux is then given by

$$q_x = -k\frac{T_1 - T_0}{h} \tag{6.8}$$

so that the temperature profile $T(x)$ can be expressed by

$$T(x) = T_1 - \frac{q_x}{k}(x - x_1) \tag{6.9}$$

It is necessary to note that specifying the boundary condition of the second kind at two spots, let say $q(0)$ at a point $x(0)$ and $q(1)$ at a point $x(1)$, would not be sufficient for the determination of the temperature profile because for $q(0)=q(1)$ there is an infinite number of solutions while for $q(0)\neq q(1)$ no steady solution exists.

6.1.3 Temperature distribution in a cylindrical layer through which heat is conducted

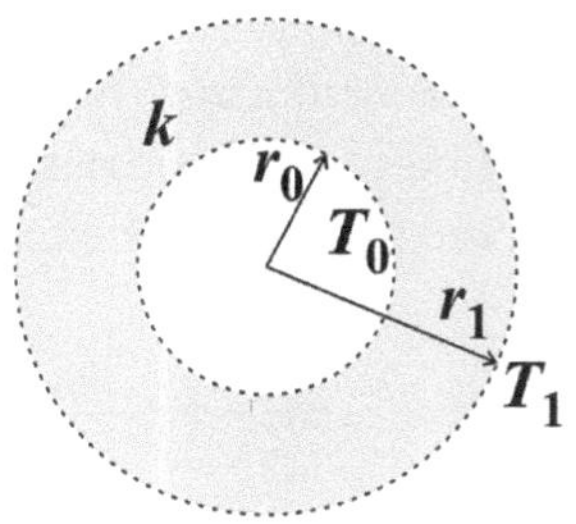

Fig. 6.2 Cross section of a cylindrical layer.

To describe the heat conduction through a cylindrical layer (Fig. 6.2), it is convenient to use cylindrical coordinates in which the temperature is constant at the constant radial coordinate r. In the differential equation it is possible to ignore the curvature of the isotherms so it holds (like for a flat plate)

$$q_r = -k\frac{dT}{dr} \tag{6.10}$$

Because the overall steady flow of heat in the radial direction

$$\frac{dQ}{dt} \equiv q_r S = q_r 2\pi r \Delta z \tag{6.11}$$

remains constant, it holds

$$\frac{dQ}{dt}\frac{1}{2\pi k \Delta z} = -r\frac{dT}{dr} \tag{6.12}$$

The integration of Eq. (6.12) results in

$$T(r) = \frac{dQ}{dt}\frac{1}{2\pi k \Delta z}\ \ln(r) + C \tag{6.13}$$

where the integration constant will be determined from the boundary condition. For example, if the temperatures T_0 and T_1 are known, then

$$T(r) = T_0 + \ln(r/r_0)\frac{T_1 - T_0}{\ln(r_1/r_0)} \tag{6.14}$$

and

$$q_r(r_0) = -k\frac{T_1 - T_0}{r_0 \ln(r_1/r_0)} \tag{6.15}$$

6.1.4 Temperature distribution in a spherical layer through which heat is conducted

In this case, provided that the parameters shown in Fig. 6.2 hold, the constant radial heat flow will be

$$\frac{dQ}{dt} = q_s 4\pi s^2 \tag{6.16}$$

so the equation for the heat conduction will be

$$\frac{dQ}{dt}\frac{1}{4\pi k} = -s^2\frac{dT}{ds} \tag{6.17}$$

The integration of Eq. (6.17) results in

$$T(s) = \frac{dQ}{dt}\frac{1}{4\pi k s} + C \tag{6.18}$$

In the next step the boundary conditions will be used as in the preceding case.

6.2 Steady conduction of heat with a source of heat

6.2.1 A flat layer with a constant source and a given outside temperature

The case considered here is a uniform heat source when each unit of volume receives in a unit of time heat energy Φ (W/m^3) at a constant temperature T_0 on both boundaries of the layer. For the sake of simplicity let us introduce a coordinate $x=0$ into the center of the layer that has a thickness $2h$ so that the coordinates of the edges of the layer are $x_0=-h$ and $x_1=h$. To a unit area ΔS, held between the planes $x=$const. and $x+\Delta x=$const., is assigned an elemental volume $\Delta V = \Delta x \Delta S$. In this volume, the amount of heat $\Delta Q = \Phi \Delta V$ increases over a unit of time so that the amount of heat coming to a unit area is $\Delta q = \Phi \Delta x$. Thus, the difference of the heat flow intensities will be

$$q(x + \Delta x) - q(x) = \Phi \Delta x \tag{6.19}$$

Converting differences to differentials gives

$$\lim_{\Delta x \to 0} \frac{q(x + \Delta x) - q(x)}{\Delta x} \equiv \frac{dq}{dx} = \phi \tag{6.20}$$

In terms of temperatures it will be

$$-k \frac{d^2 T}{dx^2} = \phi \tag{6.21}$$

Dual integration of Eq. (6.21) results in

$$T = -\frac{\phi x^2}{2k} + C_1 x + C_2 \tag{6.22}$$

After the evaluation of the integration constants from the boundary conditions the result will be a parabolic distribution of temperatures.

$$T = T_0 + \frac{\phi}{2k}\left(h^2 - x^2\right) \tag{6.23}$$

6.2.2 A cylinder with a constant source and a given outside temperature

A similar case is a uniform source of heat Φ in a cylindrical body having a temperature T_0 on its periphery, that is, at the coordinate $r = r_0$. Thus, the difference between the heat fluxes on a unit length of the elemental cylindrical layer at the coordinates r=const. and $r + \Delta r$=const. will be

$$2\pi(r + \Delta r)\, q(r + \Delta r) - 2\pi r q(r) = 2\pi r \Delta r \Phi \tag{6.24}$$

Using the same procedure as was done for a flat plate, a differential equation is obtained

$$-k \frac{d}{dr}\left(r \frac{dT}{dr}\right) = \phi r \tag{6.25}$$

By performing a double integration the expression for the temperature distribution is obtained

$$T = -\frac{\phi r^2}{4k} + C_1 \ln(r) + C_2 \tag{6.26}$$

By applying an assumption that the profile is symmetric at the axis of the cylinder ($dT/dr=0$) and by the application of the boundary conditions, an equation for the velocity profile is obtained that is again parabolic.

$$T = T_0 + \frac{\phi}{2k}\left(\frac{r_0^2}{4} - \frac{r^2}{4}\right) \tag{6.27}$$

6.3 Steady heat conduction through a composite layer

6.3.1 Serial arrangement

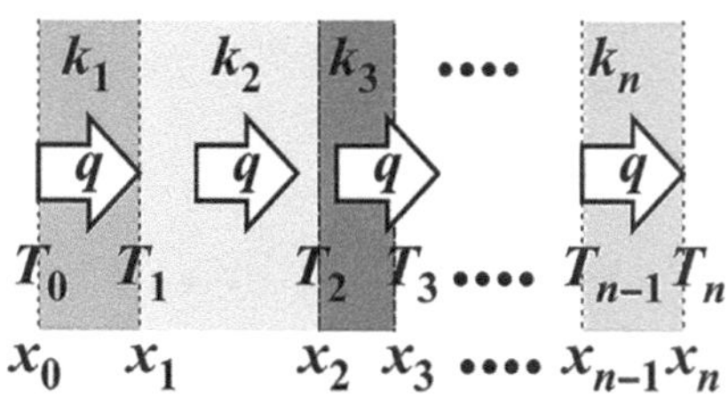

Fig. 6.3 Heat conduction through serially arranged layers.

In a material composed from several layers (Fig. 6.3), the two boundary temperatures T_0 and T_n need to be known. The intensity of flow q is the same in all layers because there is no accumulation of heat in any of the layers. Therefore the condition n

$$q = -k_i \frac{T_i - T_{i-1}}{x_i - x_{i-1}} \tag{6.28}$$

will hold for all $0 < i < n$. Therefore,

$$T_i - T_{i-1} = -q\frac{h_i}{k_i} \tag{6.29}$$

where h_i is the thickness of the "ith" layer

$$h_i \equiv x_i - x_{i-1} \tag{6.30}$$

By the summation of all equations one gets

$$T_n - T_0 = -q\sum_{i=0}^{n}\frac{h_i}{k_i} \tag{6.31}$$

The whole set of layers behaves as if it had an apparent thermal conductivity λ_{avg} meeting the condition

$$q = -k_{avg} \frac{T_n - T_0}{x_n - x_0} \tag{6.32}$$

where the apparent thermal conductivity is

$$k_{avg} = \frac{h}{\sum_{i=0}^{n} \frac{h_i}{k_i}} \tag{6.33}$$

It is obvious that the controlling role in the denominator will be played by the terms having the highest values of the ratio h_i/k_i, that is, the layers with the highest resistance. Here is an obvious analogy with the Kirchhoff's law for the conduction of the electric current through resistances serially arranged.

6.3.2 Parallel arrangement

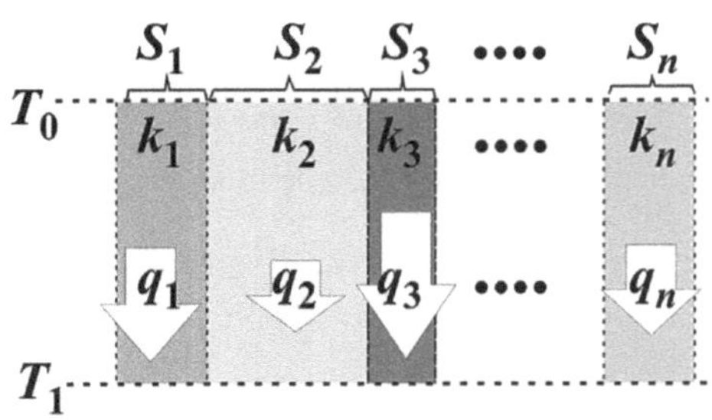

Fig. 6.4 Heat conduction through parallel layers.

If layers of miscellaneous materials are oriented at right angles toward surfaces with constant temperatures (Fig. 6.4) then there is an equal driving force $(T_0 - T_1)$ in each partial front surface S_i. It then holds

$$q_i = -k_i \frac{T_1 - T_0}{h} \tag{6.34}$$

and the overall heat flow per unit time will be

$$\frac{dQ}{dt} = \sum_{i=1}^{n} q_i S_i \tag{6.35}$$

The average heat flow for the whole layer will be

$$q_{avq} = -\frac{\sum_{i=1}^{n} k_i S_i}{\sum_{i=1}^{n} S_i} \frac{T_1 - T_0}{h} \tag{6.36}$$

so it appears as if the whole layer behaved as a uniform material having average heat conductivity

$$k_{avq} = -\frac{\sum_{i=1}^{n} k_i S_i}{\sum_{i=1}^{n} S_i} \tag{6.37}$$

It is worth noticing that in this case the controlling role is played by the materials having the highest thermal conductivity. Again, here is a clear analogy with the Kirchhoff's laws for electric current flowing through resistances arranged in parallel.

6.3.3 Heat conduction in an anisotropic medium

Certain materials are apparently homogeneous but they might have an internal structure formed by composites with layers that have diverse heat conductivities. In addition, these materials might be also anisotropic, which means that the thermal conductivity depends not only on the average composition of the material but also on the direction in which the heat moves. This can be explained by the variability of the apparent thermal conductivity according to the relationships (6.33) and (6.37). A typical anisotropy is shown by wood the properties of which along the annual rings are clearly different than those at a right angle to them. (Even more spectacular is the anisotropy of strength.) The anisotropic microstructure might not be visible at first glance.

Markedly different are the heat conductivities of graphite when measured perpendicularly to the layers of atomic size and along them.

6.4 Summary

Unidirectional steady heat conduction is described by an ordinary differential equation for the dependence of the heat flow on temperature. To obtain a complete description of the temperature distribution along the heat conduction path it is necessary to know either two local temperatures or one temperature and one value of the heat flow. If the surfaces having a constant temperature are parallel then the spatial distribution of temperatures is linear. Somewhat more complex is the case when the surfaces of constant temperature are coaxial cylindrical planes or concentric spherical planes.

The equations describing the temperature distribution can be solved even if they include a heat source.

The combination of the above solutions can be used for the description of heat conduction in layered materials. The heat flow through layers on contact arranged serially is identical. Should the outside temperatures be given then the heat flow is controlled by the layer with the highest resistance. On the contact of layers arranged in parallel with the direction of the heat flow, there is always identical temperature. Should the outside temperatures be given then the overall heat flow is primarily controlled by the layer with the least resistance.

CHAPTER SEVEN

Simple problems in fluid flows

Contents

7.1 Viscometric flows

In a number of cases the flow is controlled by the viscosity. In particularly, this is the case when the viscosity of the liquid is relatively high and the flow takes place in a restricted space. This type of flow is called the **creeping flow**. More about this type of flow will be said in Section 10.6.

Certain simple creeping flows that are used for the measurement of viscosity are denoted as **viscometric flows**. Because they occur in simple geometric setups, it is possible to assume where the liquid flows at a constant speed and in which direction the speed will change. If a suitable coordinate system is chosen, then there will be only one nonzero component of the tangential stress τ_{ij} although there will be another hidden component τ_{ji} symmetric with it. More on this subject will be given in Section 8.1. A simple calculation will suffice to determine the distribution of forces and speeds. The most common viscometric flows are the flow in a straight tube and the flow at a moving wall.

Transport and Surface Phenomena
https://doi.org/10.1016/B978-0-12-818994-8.00007-5

7.2 The flow in straight tubes

In this case the distribution of the tangential stresses can be determined from the balance of forces.

7.2.1 The tube with a circular cross section

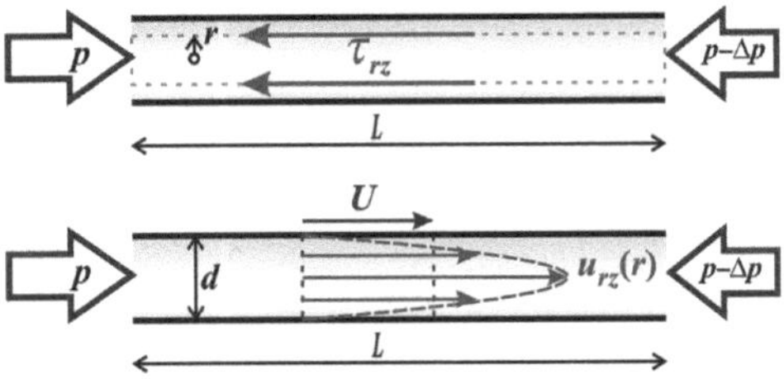

Fig. 7.1 Balance of forces and the velocity profile in a circular pipe.

Let us choose an imaginary section of a pipe having an inner diameter d being limited by a coaxial cylinder of length L and radius r. The fronts of the cylinder are exposed to pressures the difference of which is ΔP. Due to this pressure difference, a force $\pi r^2 \Delta P$ is generated that is opposed by a tangential stress τ_{zr} that has, for the sake of symmetry, the same value in the whole cylindrical plane $2\pi rL$ (see Fig. 7.1). Thus, an equality

$$\pi r^2 \Delta P = -2\pi r L \tau_{zr} \tag{7.1}$$

will hold. The tangential stress is linked to the velocity changes as determined by state function (2.35)

$$\tau_{zr} = \mu \frac{du_z}{dr} \tag{7.2}$$

The combination of these two relations will yield a simple differential equation

$$\frac{du_z}{dr} = -\frac{\Delta P r}{2\mu L} \tag{7.3}$$

the solution of which is

$$u_z = -\frac{\Delta P r^2}{4\mu L} + C \tag{7.4}$$

Considering that the velocity of the liquid is zero on the solid wall of the tube ($r = d/2$), the value of the integration constant C can be back-calculated and a parabolic velocity profile is obtained

$$u_z = \frac{\Delta P}{4\mu L}\left[(d/2)^2 - r^2\right] \tag{7.5}$$

The volumetric flow rate can be then determined by the following integration:

$$V' = \int_0^{d/2} 2\pi r u_z dr = \frac{\pi}{128}\frac{\Delta P d^4}{\mu L} \tag{7.6}$$

The mean velocity of the flow can be then calculated from the volumetric flow rate by

$$U = \frac{V'}{\pi d^2/4} = \frac{1}{32}\frac{\Delta P d^2}{\mu L} \tag{7.7}$$

Eq. (7.7) is used to evaluate the viscosity of a liquid from experimentally determined flow rates and pressures in a capillary viscometer (e.g., Fig. 7.2).

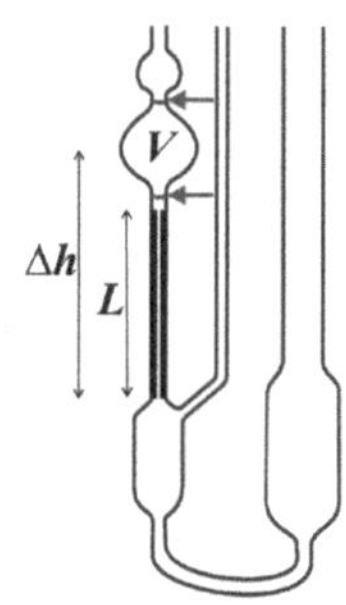

Fig. 7.2 Ubbelohde type capillary viscometer. The capillary with a measuring bulb is on the left, other piping is auxiliary.

7.2.2 Noncircular geometries

In addition to the flow in a circular pipe, two more unidirectional flow situations can be solved. They are the flow in a flat channel and the flow in a pipe of annular cross section where a circular tube is placed coaxially in another circular tube. In both these cases it is assumed that the velocity at the wall is zero.

This flow situation is met at the extrusion of polymers through nozzles while manufacturing hosepipes or when extruding dough to make macaronis.

Another flow situation that can be solved is the case when the central cylinder in an annular flow moves in the z direction.

A good example of this flow is the coating of wires by polymers. Usually, the polymer is extruded through an annular nozzle so that there is a zero velocity gradient of the polymer melt at the surface of the wire.

This subject will be dealt with again in Section 12.1.

7.3 Rotary flow along a moving boundary

This flow situation is one of the so-called Couette flows. An interesting case is the unidirectional longitudinal flow depending on only one diagonal coordinate. In this case the velocity distribution can be calculated.

As long as the gap between the two cylindrical planes is narrow, the curvature can be neglected and the flow can be treated in Cartesian coordinates. The longitudinal coordinate will be denoted by x and the diagonal by y.

7.3.1 Rotational viscometer with two coaxial cylinders

The flow between two infinite flat plates, used to define the viscosity, cannot be physically realized. Instead, the layer of the liquid is realized between two coaxial rotating cylinders. The simplest realization of a **rotational viscometer** is a temperature controlled cylindrical container in which a spindle rotates at a given or measured angular velocity ω while the torque M is measured or set (see Fig. 7.3).

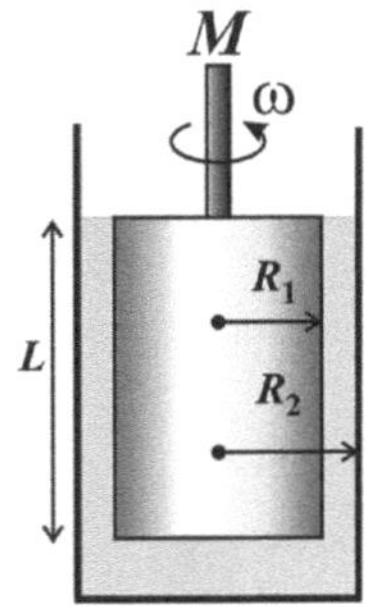

Fig. 7.3 Coaxial cylinders viscometer geometry.

An important advantage of this viscometer is that it needs only a small amount of the measured sample. By placing the liquid level close to the upper edge of the rotor and an air bubble underneath the lower front of the rotor, it will make it possible to neglect whatever happens at the fronts of the rotor. Then it is possible to consider the rotor to be a section of an infinitely high cylindrical area. Alternatively, the influence of the fronts of the rotor can be estimated from the experimental determination of the torque moments when rotors of different length L_2 and L_1 are used. The difference of the torque moments $M_2 - M_1$ will depend only on the difference of the lengths $L_2 - L_1$ of the rotors. If the setup of two cylinders can be viewed to be a section of an infinite cylinder having length L, then the velocity field in the gap between the two cylinders will be characterized by only the angular velocity $\omega(r)$ and single function $\tau_{r\varphi}(r)$. This is the standard case of the Couette flow. Again, the shear rate is the ratio of the relative velocity of two neighboring layers and their distance. In the limit, it is equal to $\gamma = r\, d\omega/dr$.

It is worth noticing that the shear rate in this case differs from the velocity gradient that is equal to $dv_\varphi/dr = d(\omega r)/dr$. It would acquire a nonzero value $dv_\varphi/dr = \omega$ even if the liquid is rotated as a solid body without being subject to any shearing. This subject will be discussed again in the section on multidimensional flow.

The basic relationship for a given shear flow is

$$\tau_{r\varphi} = \mu r \frac{d\omega}{dr} \tag{7.8}$$

The liquid transfers the torque momentum M that is independent of r. Thus, the local tangential stress is

$$\tau_{r\varphi} = \frac{-M}{2\pi r^2 L} \tag{7.9}$$

By combining (7.8) and (7.9) one gets a differential equation

$$\frac{d\omega}{dr} = \frac{-M}{2\pi\mu r^3 L} \tag{7.10}$$

The solution of which for the following boundary conditions

$$\omega = \Omega \text{ for } r = R_1 \tag{7.11}$$

$$\omega = 0 \text{ for } r = R_2 \tag{7.12}$$

is the velocity profile

$$\omega = \Omega \frac{\frac{1}{r^2} - \frac{1}{R_2^2}}{\frac{1}{R_1^2} - \frac{1}{R_2^2}} \tag{7.13}$$

and the torque momentum

$$M = \frac{4\pi\mu\Omega L}{\dfrac{1}{R_1^2} - \dfrac{1}{R_2^2}} \tag{7.14}$$

The experimental conditions can be easily varied, for example, by varying the frequency of rotation of the rotor. If each measurement yields the same viscosity

$$\mu = \frac{M}{4\pi L\Omega}\left(\frac{1}{R_1^2} - \frac{1}{R_2^2}\right) \tag{7.15}$$

then the liquid follows Newton's law. In other cases, it belongs to non-Newtonian materials.

7.3.2 The case of a narrow gap

If the gap between the two cylinders $h \equiv R_2 - R_1$ is very narrow, that is, if $h \ll R_1$, then relationship (7.15) can be expanded using the Taylor development in a series the first two members of which are

$$M = \frac{2\pi\mu\Omega L R_1^3}{h}\left(1 - \frac{6\,h}{R_1} + \ldots\right) \approx \frac{2\pi\mu\Omega L R_1^3}{h} \tag{7.16}$$

Then the tangential stress at the wall R_1 should be

$$\tau_W = \frac{M}{2\pi L R_1^2} \approx \mu\frac{\Omega R_1}{h} \tag{7.17}$$

Because the product ΩR_1 is the peripheral speed at the wall of the rotating cylinder and because the expression $(\Omega R_1/\mathrm{h})$ approaches in the limit the shear rate in the entire space, this result is essentially identical with the definition of the viscosity as given in Eq. (3.8). Consequently, the viscometers with two concentric cylinders forming a very narrow gap are very popular.

7.3.3 Other rotational viscometric flows

Just like the flow between two coaxial cylinders, there are other simple geometric configurations in which the velocity distribution has a high regularity (Fig. 7.4) and which can be used to measure the viscosity. Specifically, these

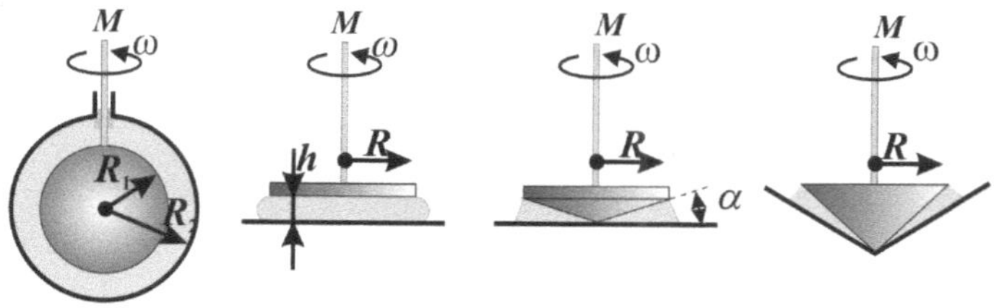

Fig. 7.4 Rotation viscometer configurations: double sphere; double disc; cone and plate; double cone.

are geometric setups having a very narrow gap between the rotor and the stator. Among them are those where

- the rotor is a flat disc and the stator is a flat plate. In such a setup combining **two flat plates**, the individual planar liquid layers move at a constant angular velocity that depends only on the distance from the rotor and the stator;
- the rotor is a cone and the stator is either a flat plate or also a cone. In such a setup, a **cone-cone** or a **cone-flat plate**, the individual conical layers of the liquid move at a constant angular velocity that depends only on the angle between the rotor and the stator. Spherical coordinates are suitable for the description of such a flow;
- both the rotor and the stator are spheres. Such a setup of a **sphere-sphere** would be inconvenient both to make and to operate. For a Newtonian liquid it is to be expected that the individual liquid layers occupying concentric space would move at a constant angular velocity. This flow setup will be dealt with in Section 12.3.

While using the rotational viscometers (including the one with a rotating cylinder) it is convenient to use instruments with a very thin layer of liquid; for once, a thin layer will have more uniform distribution of shear rates, and further there is a greater guarantee that the flow will be in the creeping flow region even when measuring liquids of low viscosity. Also, there would be a lower risk of formation of whirls in the liquid. And further, the measurement carried out in a narrow gap will require a higher torque and thus more accurate measurement.

7.4 Measurement of viscosity by nonviscometric setups

How to use nonviscometric flows for the measurement of viscosity?

In general, viscosity can be determined from the ratio of a dynamic quantity (pressure loss, hydraulic head, torque, etc.) and a kinematic quantity (speed, flow rate, etc.) in the same way as it is done when using geometrically simple flows

$$\mu = k \times \frac{\text{dynamic quantity}}{\text{kinematic quantity}} \tag{7.18}$$

where k is a custom dimension constant that can be obtained for any flow situation in which a creeping flow can be realized (i.e., a flow controlled by viscosity). This can be done either by a demanding solution of more complex equations or by a calibration using a liquid of known viscosity.

7.4.1 The case of a rotor in a large space

In the case when the cylinder moves in a large container, $R_2 \gg R_1$, it can be found that the wall shear stress at the wall of an "infinitely long" cylinder will be

$$\tau_W \approx 2\mu\Omega \tag{7.19}$$

For the **immersion rotating viscometers**, the rotors of which can be of more general geometry than coaxial cylinders, neither the shear rate at the wall nor the tangential stress can be exactly defined since the flow in such a setup is no longer viscometric. Yet even for an arbitrary spindle (like the one in Fig. 7.5), here the viscosity can be determined from the measured torque moment and the velocity provided that the viscometer operates in the creeping regime of flow and the liquid is a Newtonian one. Thus,

$$\mu = k\frac{M}{\Omega} \tag{7.20}$$

where k is a constant determined by calibration with a liquid of known viscosity.

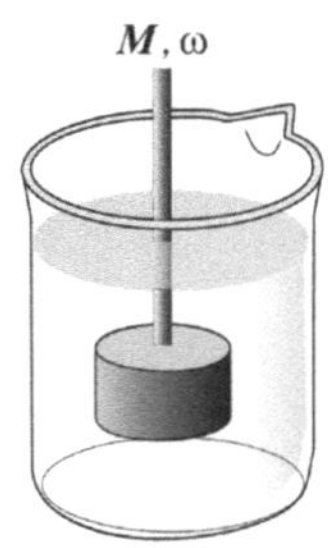

Fig. 7.5 Spindle of an immersion rotating viscometer.

7.4.2 The discharge viscometers

A very simple viscometer consists of a container from which a liquid discharges through a well-defined quantity in the time needed for a set amount of the liquid to flow out. The viscosity is then determined from

$$\mu = k \times \text{duration of the outflow} \times \text{density of the liquid} \qquad (7.21)$$

7.4.3 Viscometers with a falling body

Another type of viscometer using the weight of a standardized body falling in the measured liquid as the dynamic variable is an instrument in which the measured quantity is the time needed to cover a set distance

$$\begin{aligned}\mu = {} & k \times \text{time for the fall} \\ & \times (\text{density of the liquid} - \text{density of the liquid})\end{aligned} \qquad (7.22)$$

The constant k for a sphere falling in a free space can be determined from the Stokes law. However, it is more convenient to measure the time for the fall in a limited space because one does not need a large volume of the liquid sample and because it is easier to maintain a constant temperature of the sample. A frequently used viscometer of this type is the Hoppler viscometer with a ball falling and partially also rolling in a sloping tube (see Fig. 7.6).

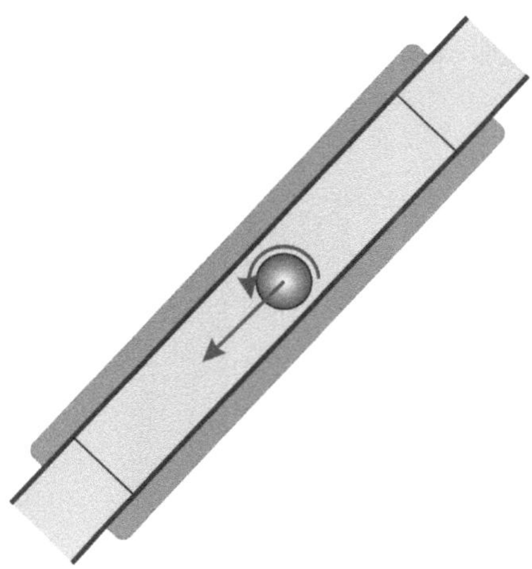

Fig. 7.6 Flow in a sloping tube of the Hoppler viscometer.

The constant k, that is determined by calibration, is usually contained in the documentation provided by the manufacturer.

7.5 Summary

The knowledge of the distribution of velocities and stresses can be obtained by elementary mathematical tools only if there is one direction, expressed in simple coordinate systems (Cartesian, cylindrical, or spherical), in which the velocity can change. Such an arrangement is called a viscometric flow since it can be used for the determination of the absolute value of viscosity. In addition, viscometric flows can be used to measure the flow properties of the non-Newtonian liquids. The quantities needed for the determination of viscosity that can be directly measured are kinematic quantities (such as the local velocities or flow rates) and the dynamic variables (such as the power input, pressure loss, or the torque). One of these variables is set and the other is measured.

The capillary viscometers are based on the use of tubes having circular cross section although, in principle, a flat slit or an annular cross section can be also used.

The rotational viscometers are based on the measurement of the torque moment while the liquid rotates between two coaxial cylinders or two concentric spheres. Similar are setups with a narrow slit between two rotating discs or between a cone and a disc.

There are other options of how to measure the viscosity of liquids that use well-defined geometric configurations. However, these are only comparative measurements requiring calibration with liquids of a known viscosity.

CHAPTER EIGHT

Fundamentals of the fluid mechanics

Contents

8.1 Plane forces, stress tensor, and its components

For complex space flows, **plane forces** can be identified that act on the surface of a fluid element. These forces can be described by a **stress tensor** $\boldsymbol{\sigma}$ the components of which σ_{ij} represent the components of a force acting from the outside in the i direction on a plane oriented in the j direction.

Tensor (see Section 21.3) is a mathematical entity relating two vectors. In a three-dimensional space it has nine elements. In a handwritten text it is marked with two arrows on top of the symbol $\vec{\vec{\sigma}}$. *However, block or highlighted letters* $\boldsymbol{\sigma}$ *are preferred in a typeset text.*

The **diagonal components** of the stress tensor express the **normal stresses** that have a nonzero value also in the case when the fluid is not deformed. In hydrostatics, it is customary to consider these components to be equivalent to **negative pressure**

$$\sigma_{ii} = -p \tag{8.1}$$

The minus sign follows from the convention that the tensor $\boldsymbol{\sigma}_{ii}$ is positive if the force acts on the plane from the outside. This follows from the general

Transport and Surface Phenomena
https://doi.org/10.1016/B978-0-12-818994-8.00008-7

habit used in the study of elasticity that positive is the pulling force. The average value of the diagonal components of the stress tensor is defined in hydrodynamics as the **isotropic pressure**

$$p \equiv -\frac{1}{3}\sum_i \sigma_{ii} \tag{8.2}$$

In a compressible fluid, the pressure can manifest itself by influencing the density as predicted by the equation of state but it has no influence on the shear deformation. If the pressure component is removed from the stress tensor $\boldsymbol{\sigma}$, then what is left is the **tensor $\boldsymbol{\tau}$ of dynamic stresses (extra stress tensor) $\boldsymbol{\tau}$**

$$\boldsymbol{\tau} \equiv \boldsymbol{\sigma} - p\boldsymbol{I} \tag{8.3}$$

where $\boldsymbol{I}$ is the diagonal unit tensor.

The tensor of dynamic stresses is associated with forces due to the deformation. Specifically, in solid bodies, all components are significant while in fluids only the tangential components play a significant role due to the manifestation of the **viscous friction.**

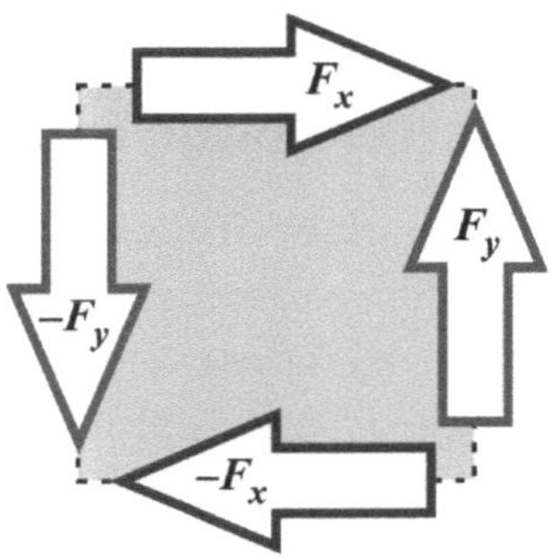

Fig. 8.1 Balance of the tangential forces.

An important feature of the stress tensor is its **symmetry** (see Fig. 8.1). It is not obvious in the case of a simple shear flow used to define the viscosity. In this case, a **tangential stress** τ_{yx} is considered to be acting on a small element of fluid from one side being compensated by a reacting stress $-\tau_{yx}$ from the opposite side. According to classical physics, such a couple of forces would result in the rotation of the fluid element. Therefore, the shear flow must involve some other forces (obvious or hidden) that prevent the element from rotation. Also, the element of the fluid must be always subject to the tangential components of the stress tensor $\tau_{ij} = \tau_{ji}$. Thus, by their physical nature, both the tensors $\boldsymbol{\tau}$ as well as $\boldsymbol{\sigma}$ must be symmetrical.

8.2 Three-dimensional generalization of Newton's law of viscosity

So far, Newton's law of viscosity was presented only for the case of a simple shear flow with one velocity component u_y, varying in one direction $x \neq y$ only

$$\tau_{yx} = \mu \gamma \tag{8.4}$$

For quantitative considerations involving velocities and forces (i.e., in kinematics and dynamics) in multidimensional flow of a continuum, however, it is necessary to know the relationships between all components of the tensor of dynamic stresses $\boldsymbol{\tau}$ and of shear rate that depends on the **kinematic tensor** $\nabla \boldsymbol{u}$.

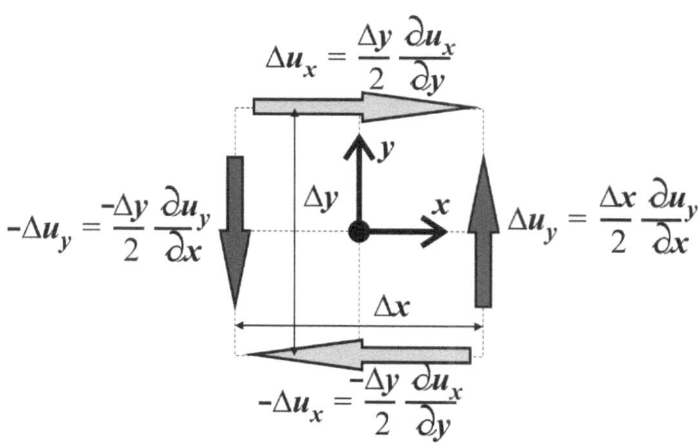

Fig. 8.2 Elements of the kinematic tensor in the *x-y* plane.

Its components (see Fig. 8.2) are the elements of the matrix expressed as:

$$\nabla \boldsymbol{u} \equiv \left\| \frac{\partial u_i}{\partial j} \right\|_{yx} \tag{8.5}$$

where the indices i, j may, in the Cartesian coordinate system, acquire values x, y, z.

It appears, however, that there is a complication that does not occur in the case of the relationship between the vector depicting the intensity of the heat flow and the temperature gradient. Some of the nine components of the velocity gradient are not linked to the shear deformation. Neither is stress tensor linked in a simple way to this deformation.

8.2.1 The role of the kinematics of the fluid on the deformation

The rotation of a block of a fluid does not contribute to its shear deformation. For the rotation it holds

$$\frac{\partial u_y}{\partial x} = -\frac{\partial u_x}{\partial y} \tag{8.6}$$

Though there will be a nonzero velocity gradient when the body rotates, a deformation does not occur. Therefore, an antisymmetric **vorticity tensor** will be introduced by a relationship

$$\boldsymbol{\Omega} \equiv \left\| \frac{1}{2}\left(\frac{\partial u_i}{\partial j} - \frac{\partial u_j}{\partial i}\right) \right\| = \frac{1}{2}(\nabla \boldsymbol{u} - \boldsymbol{u}\nabla) \tag{8.7}$$

the components of which cohere with the rotation but do not cohere with shear deformation (see Fig. 8.3).

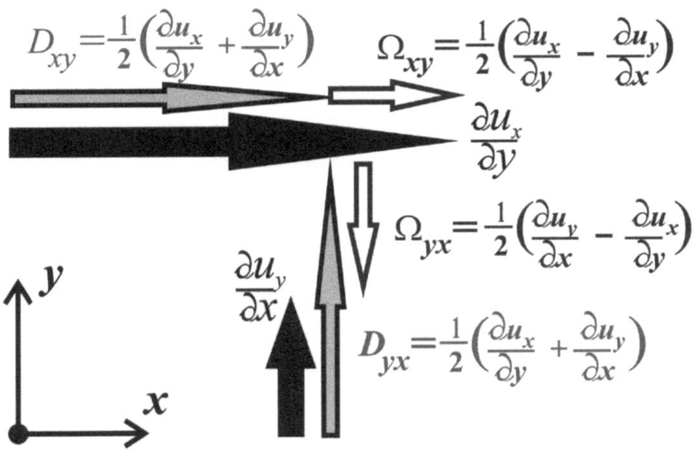

Fig. 8.3 Partial differentials of the components of tangential velocities. The real values in the z=const. plane are in black. The average values belong to $\underline{\boldsymbol{D}}$ that determine the deformation. One-half of the difference belongs to $\underline{\boldsymbol{\Omega}}$ that determines the rotation.

It is possible to separate the rotation from the kinematics of deformation by keeping only the symmetric part of the kinematic tensor. Further, it is customary to introduce a **tensor of deformation rates**

$$\boldsymbol{D} \equiv \left\| \frac{1}{2}\left(\frac{\partial u_i}{\partial j} + \frac{\partial u_j}{\partial i}\right) \right\| = \frac{1}{2}(\nabla \boldsymbol{u} + \boldsymbol{u}\nabla) = \nabla \boldsymbol{u} - \boldsymbol{\Omega} \tag{8.8}$$

which characterizes the deformation.

Similarly, as it is with the components of the deformation tensor also, this tensor has only six independent components.

For a simple shear flow oriented in any general direction, the **shear rate** is equal to the double of the square root of the double of the scalar product of the tensors (see definition 21.33). The shear rate is equal to the velocity gradient only if the flow is non-whirling ($\boldsymbol{\Omega}=0$), otherwise

$$\gamma = 2\sqrt{\mathbf{D}:\mathbf{D}} \tag{8.9}$$

In theory, the shear rate could depend also on all scalar invariants (21.34 to 21.36) but the first invariant is equal to zero by definition (10.3) and the standard liquids are independent on the third invariant.

Example:

A simple shear flow rotated by an angle α in the x,y plane (Fig. 8.4).

It is assumed for the above flow, without placing it in an x,y coordinate system, that the only significant component is

$$\gamma = \frac{\Delta U}{h}$$

If this flow is, however, treated in this coordinate system, then the distance between the two plates is

$$\Delta x = h \sin \alpha$$
$$\Delta y = -h \cos \alpha$$

and for the two velocity components

$$\Delta u_x = \Delta U \cos \alpha$$
$$\Delta u_y = -\Delta U \sin \alpha$$

Therefore,

$$2\sqrt{\mathbf{D}:\mathbf{D}} = \frac{\Delta U}{h} = \gamma \tag{8.10}$$

which is the expression assumed by relationship (8.9).

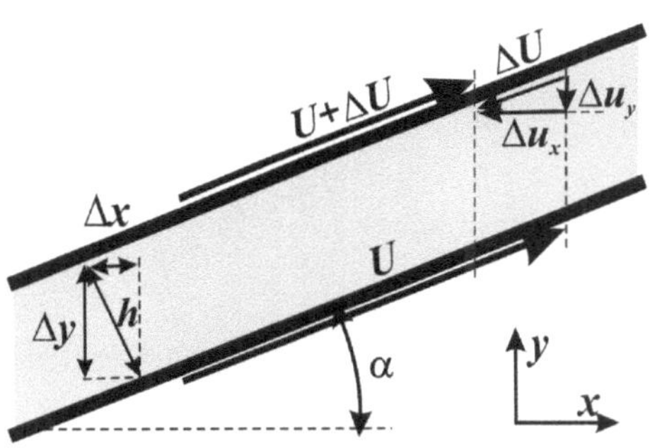

Fig. 8.4 Shear flow in arbitrary orientation.

8.3 Rheology

Rheology is a scientific discipline dealing with the deformation and flow of materials. It does it from several different positions:

Experimental rheology-rheometry strives primarily at testing materials: these are, for example, extension tests of solid materials and viscometric studies for liquids. There are, however, a number of materials important in engineering practice that have a more complex flow behavior. For these it is necessary to develop specialized methods and instruments.

Theoretical rheology strives to obtain mathematical expressions for the measured rheological properties. The aim here is to apply these expressions in the **mechanics of continua** to obtain mathematical tools by which the rheological behavior could be interpreted and predicted under general conditions.

Structural rheology seeks to find the link between the rheological behavior and the structure of the material. For simple liquids their viscosity could be approximately predicted if the sizes of the molecules and the intermolecular forces are known. Conversely, the mean molecular mass of macromolecules can be estimated from the systematic measurement of the viscosity of polymer solutions.

8.3.1 The rheological constitutive equation of liquids

It is assumed that there is a unique assignment of the quantities $\boldsymbol{\tau}$ and $\boldsymbol{D}$, which characterizes the **rheological behavior** of a fluid, that is a relationship between the dynamics and the kinematics of the deformed fluid. A mathematical expression of this relationship is known as the **rheological constitutive** equation. This equation makes it possible to calculate the rheological behavior of the fluid exposed to a set of given forces and, conversely, what forces are needed to perform a required motion.

The simplest, and also the most common, is the constitutive equation of the **Newtonian liquid,** which just generalizes the definition of **viscosity** for the spatial flow. Thus, it holds

$$\boldsymbol{\tau} = 2\mu \boldsymbol{D} \tag{8.11}$$

or

$$\tau_{ij} = \tau_{ji} = \mu\left(\frac{\partial u_i}{\partial j} + \frac{\partial u_j}{\partial i}\right) \tag{8.12}$$

8.3.2 The energy dissipation by viscous friction

Viscous friction results in an irreversible conversion of the mechanical energy in heat (scalar quantity). The **intensity of dissipation** ε_{dis} that is the amount of the mechanical energy converted to heat in a unit volume in a simple shear flow is equal to a product

$$\varepsilon_{dis} = \tau_{yx} \frac{\partial u_x}{\partial y} \tag{8.13}$$

It can be derived that the intensity of dissipation for a spatial flow is equal to a double scalar product (see definition 21.33)

$$\varepsilon_{dis} = \boldsymbol{\tau} : \boldsymbol{D} \tag{8.14}$$

For a Newtonian liquid the following relationship can be used

$$\varepsilon_{dis} = (\boldsymbol{\tau} : \boldsymbol{\tau})/(2\mu) = 2\mu(\boldsymbol{D} : \boldsymbol{D}) = \mu\gamma^2 \tag{8.15}$$

which reduces for a one-dimensional flow to

$$\varepsilon_{dis} = \boldsymbol{\tau} : \boldsymbol{D} \tag{8.16}$$

The intensity of dissipation plays a significant role in the polymer processing industry where the heat dissipated in the heavy kneading machines is sufficient to heat the processed polymer. For most of the liquid processing operations, however, the temperature increase is insignificant and it is seen merely as a **loss of the mechanical energy**.

8.3.3 Non-Newtonian liquids

Among the polymer solutions and melts, suspensions and pastes, emulsions and creams, and other liquid and semiliquid materials that frequently occur in many technologies of the chemical and processing industries, there are many liquids for which the relationship between the tensors $\boldsymbol{\tau}$ and $\boldsymbol{D}$ is unique but is not linear (see Fig. 8.5). These materials are termed **Non-Newtonian liquids**.

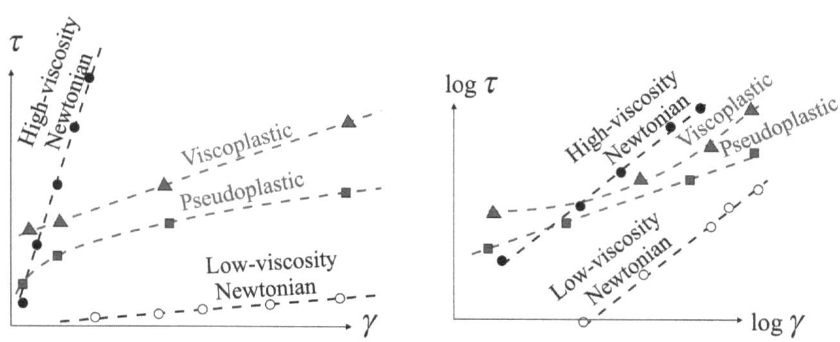

Fig. 8.5 Example of Newtonian and non-Newtonian liquid rheogram in linear and logarithmic coordinates.

The viscosity of the simplest non-Newtonian liquids, known as the **generalized Newtonian liquids**, depends, apart from the estate variables, also on the mechanical straining of the liquid, which can be expressed as the intensity of dissipation, ε. Among these the simplest one follows the so-called **power-law** model for which, assuming a simple shear flow, relationship (8.17) applies

$$\tau_{yx} = K\left|\frac{du_y}{dx}\right|^{n-1}\frac{du_y}{dx} = K\gamma^n \tag{8.17}$$

For a three-dimensional flow, relationship (8.17) applies

$$\boldsymbol{\tau} = 2K\left|2\boldsymbol{D}:\boldsymbol{D}\right|^{\frac{n-1}{2}}\boldsymbol{D} = 2K\gamma^{n-1}\boldsymbol{D} \tag{8.18}$$

The group $K\gamma^{n-1}$ is the **apparent viscosity**. The power-law model suits a large group of liquids. In most cases the apparent viscosity drops with the applied dynamic stress, which means that for them the **flow index** $n<1$. For a limiting case when $n=1$ the power law describes a Newtonian liquid for which the **consistency coefficient** $K=\mu$ becomes the viscosity. There are many more mathematical models of non-Newtonian liquids fitting better or worse the real behavior.

Plastic materials show a **yield stress** meaning that they begin to flow only after the yield stress is overcome (similar to solid materials overcoming the limit of flexibility). Typical examples of this behavior are shown by pastes, creams, and mashed suspensions.

A complete description of the behavior of the non-Newtonian liquids is, however, beyond the scope of this work. The **viscoelastic liquids** will remember the previous state. They will partially return to it when the stress is relieved. Polymer solutions and melts belong in this category. To describe the relationship between the stress and the deformation for these liquids it is necessary to choose models that have the character of differential or integral equation involving time as one of the variables

Thixotropic materials become thinner after the stress is applied for a certain time (paints, yogurt). Thus, time must be added to the rheological model.

8.4 Summary

A significant role in transport processes is played by the tangential forces that act in all directions and that are dependent on the orientation of the planes on which they act. Therefore, it is necessary to express them by all the nine components that form together a mathematical object called the tensor.

The normal components of the stress tensor (lying in the diagonal) are in regular fluids given only by the pressure that, in an incompressible medium, does not participate in the process of deformation. After the influence of the pressure from the stress tensor is eliminated, all what is left is the tensor of dynamic stresses expressing the influence of the viscous friction. Because the fluid in elementary volumes does not independently rotate, there must be (usually hidden) another tangential force so that this tensor must be symmetric. At the most, only three components are needed to describe it. While, following Newton's Law of viscosity for unidirectional flow, the only observable component of the tangential stress depends on the velocity gradient, in the multidirectional flows also the shear rate comes in the picture. This is the combination of the components of local velocity changes comprehensibly described by the tensor of deformation rates.

For Newtonian fluids, that is for the common liquids and gases, there is a direct proportionality between the tensor of dynamic stresses and the tensor of deformation rates while the dynamic viscosity is the constant of proportionality.

For the non-Newtonian fluids, however, the viscosity depends not only on the state variables but also on the flow situation. Rheology is the scientific discipline dealing with these liquids. There are, however, also liquids that display properties of solids such as elasticity and plasticity. The rheology of such materials is significantly more complicated.

CHAPTER NINE

Transports in moving media

Contents

9.1 Mass balances—The equation of continuity

The mass is a quantity for which applies the law of conservation. Verbally the mass balance can be expressed by the following statement:

$$\textit{The accumulation of mass} + \textit{the outflow of mass} - \textit{the inflow of mass} = 0 \quad (9.1)$$

9.1.1 The detailed derivation of this statement

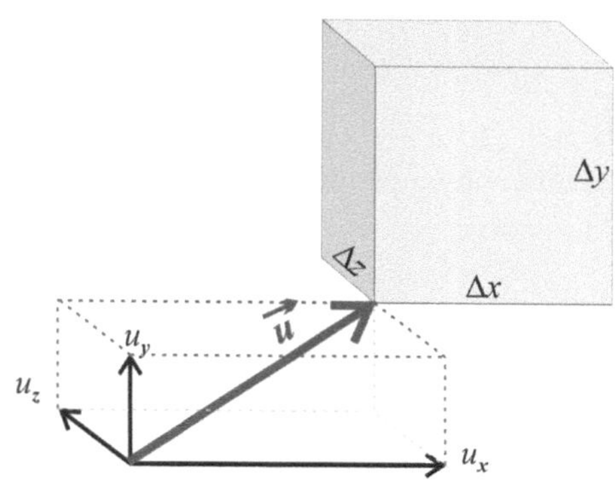

Fig. 9.1 The volume element and the velocity vector.

Transport and Surface Phenomena
https://doi.org/10.1016/B978-0-12-818994-8.00009-9

In a volume element (Fig. 9.1), $\Delta V = \Delta x\ \Delta y\ \Delta z$ of a continuum, positioned at the coordinate (x, y, z) of a Cartesian coordinate system, the accumulation of mass is equal to the difference of the final amount of mass

$$\rho(x, y, z, t+\Delta t)\ \Delta x \Delta y \Delta z$$

minus the initial amount of mass

$$\rho(x, y, z, t)\ \Delta x \Delta y \Delta z$$

If a liquid passes through this element at a speed $\boldsymbol{u}$ then the mass that enters with the liquid over the time period Δt from the region of lower x values will be

$$\rho(x, y, z, t)\ u_x(x, y, z, t)\ \Delta y \Delta z \Delta t$$

while the inflow into the region of higher x values will be

$$\rho(x+\Delta x, y, z, t)\ u_x(x+\Delta x, y, z, t)\ \Delta y \Delta z \Delta t$$

Analog expressions hold for the y and z directions.

In summary, the mass balance yields a relationship

$$\begin{array}{c}
+\rho(x, y, z, t+\Delta t)\ \Delta x \Delta y \Delta z \\
-\rho(x, y, z, t)\ \Delta x \Delta y \Delta z \\
+\rho(x+\Delta x, y, z, t)\ u_x(x+\Delta x, y, z, t)\ \Delta y \Delta z \Delta t \\
+\rho(x, y+\Delta y, z, t)\ u_y(x, y+\Delta y, z, t)\ \Delta y \Delta z \Delta t \\
+\rho(x, y, z+\Delta z, t)\ u_z(x, y, z+\Delta z, t)\ \Delta y \Delta z \Delta t \\
-\rho(x, y, z, t)\ u_x(x, y, z, t)\ \Delta y \Delta z \Delta t \\
-\rho(x, y, z, t)\ u_y(x, y, z, t)\ \Delta x \Delta z \Delta t \\
\underline{-\rho(x, y, z, t)\ u_z(x, y, z, t)\ \Delta x \Delta y \Delta t} \\
=0
\end{array}$$

9.1.2 The result

After dividing all members by a product ($\Delta x\ \Delta y\ \Delta z\ \Delta t$) and making use of the definition of a derivative

$$\frac{\partial f(x)}{\partial x} \equiv \lim_{\Delta x \to 0} \frac{f(x+\Delta x) - f(x)}{\Delta x} \tag{9.2}$$

one gets a differential relationship

$$\frac{\partial \rho}{\partial t} + \frac{\partial(\rho u_x)}{\partial x} + \frac{\partial(\rho u_y)}{\partial y} + \frac{\partial(\rho u_z)}{\partial z} = 0 \tag{9.3}$$

that can be also written in a vector form as

$$\frac{\partial \rho}{\partial t} + \vec{\nabla} \cdot (\rho \vec{u}) = 0 \tag{9.4}$$

Bold letters will be used in this text for vectors and tensors while a dot will be used for the scalar products of vectors

$$\frac{\partial \rho}{\partial t} + \boldsymbol{\nabla} \cdot (\rho \boldsymbol{u}) = 0 \tag{9.5}$$

This relationship is known as the **continuity equation**.

In many cases occurring in the hydrodynamics of fluids and even in the mechanics of gases, the influence of density variation on flow is often negligible. Then the **continuity equation of an incompressible fluid** will assume a simplified form

$$\boldsymbol{\nabla} \cdot \boldsymbol{u} = 0 \; (\text{for } \rho \to \textit{const.}) \tag{9.6}$$

9.2 Momentum balances in terms of stresses: Cauchy equation

Newton introduced a quantity that he called "the amount of motion." The contemporary term for this quantity is **momentum**.

The momentum of a material point is the product of mass m and velocity $\boldsymbol{u}$. The second Newton's law states that the change of momentum of a material point with time is given by the sum of all forces acting on this point

$$\frac{\partial (m\boldsymbol{u})}{\partial t} = \sum \boldsymbol{F}_i \tag{9.7}$$

The accumulation of momentum in a volume element of the continuum is given on the one hand by the inflow and the outflow of the momentum, and on the other hand by the action of all forces on the element. Similarly, as was the case with the balance of mass, it will hold for an element through which a liquid flows at a speed $\boldsymbol{u}$ that the amount of momentum arriving with the flowing liquid over a time period Δt from the region of smaller values of x

$$\rho(x, y, z, t)\, \boldsymbol{u}(x, y, z, t)\, u_x(x, y, z, t)\, \Delta y \Delta z \Delta t$$

After adding and deducting all the relevant terms and after dividing the result with the group ($\Delta x\ \Delta y\ \Delta z\ \Delta t$) the **accumulation of momentum** related to unit differential volume and to a unit differential time is obtained as

$$\frac{\partial(\rho \boldsymbol{u})}{\partial t} = \rho\frac{\partial \boldsymbol{u}}{\partial t} + \boldsymbol{u}\frac{\partial \rho}{\partial t} \tag{9.8}$$

Similarly, the difference between the outflow and inflow of momentum in the flowing fluid will be

$$\boldsymbol{u}\cdot\nabla(\rho\boldsymbol{u}) = \rho\boldsymbol{u}\cdot\nabla\boldsymbol{u} + \boldsymbol{u}\cdot\boldsymbol{u}\nabla\rho \tag{9.9}$$

The sum of the accumulation and of the difference of the outflow and the inflow can be regrouped as

$$\rho\frac{\partial \boldsymbol{u}}{\partial t} + \rho\boldsymbol{u}\cdot\nabla\boldsymbol{u} + \boldsymbol{u}\cdot\left(\frac{\partial\rho}{\partial t} + \boldsymbol{u}\nabla\rho\right) \tag{9.10}$$

where the term in the brackets is, according to the equation of continuity, equal to zero so that the **change of that momentum** of a unit element in a unit of time can be expressed by a quantity

$$\rho\left(\frac{\partial \boldsymbol{u}}{\partial t} + \boldsymbol{u}\cdot\nabla\boldsymbol{u}\right)$$

According to Newton's second law of motion, the change of momentum is equal to the sum of all forces acting on the element. There are only two kinds of forces acting on a unit element of volume:

Volume forces acting on a volume element are usually those due to the gravity $\boldsymbol{g}$. (In special cases, e.g., the magnetic force may act as well.)

Flat forces contribute to the change of momentum of a unit differential volume by the set of all forces acting on its surface, that is, $\nabla\cdot\boldsymbol{\sigma}$.

In general, the momentum balance is represented by the following equation:

$$\rho\left(\frac{\partial \boldsymbol{u}}{\partial t} + \boldsymbol{u}\cdot\nabla\boldsymbol{u}\right) = \rho\boldsymbol{g} + \nabla\cdot\boldsymbol{\sigma} \tag{9.11}$$

By expanding and combining with Eq. (8.3), Eq. (12.11) becomes

$$\rho\left(\frac{\partial \boldsymbol{u}}{\partial t} + \boldsymbol{u}\cdot\nabla\boldsymbol{u}\right) = \rho\boldsymbol{g} - \nabla p + \nabla\cdot\boldsymbol{\tau} \tag{9.12}$$

Usually, it is possible to find an expression for the **potential of pressure and of volume forces** P that meets the relationship

$$\nabla P \equiv \nabla p - \rho\boldsymbol{g} \tag{9.13}$$

If it holds simultaneously that $\boldsymbol{u}=0$ and $\boldsymbol{\tau}=0$, then $P=\text{const}$. This statement expresses the well-known experience that in an immobile liquid in a gravitational field, the pressure drops proportionately with the growing depth (Pascal's law). Then, it can be written as

$$\rho\left(\frac{\partial \boldsymbol{u}}{\partial t} + \boldsymbol{u}\cdot\nabla\boldsymbol{u}\right) = -\nabla P + \nabla\cdot\boldsymbol{\tau} \tag{9.14}$$

which is the common form of **Cauchy's equation of momentum** in terms of stresses. This equation, containing three unknown dependent quantities, namely $\boldsymbol{u}$, P, and $\boldsymbol{\tau}$, can be solved if it can be assumed, on the basis of the force balance in systems with high geometrical symmetry, a simple distribution of the key components of the stress tensor; there are a few such systems. Their solution is easy and it will be discussed in Section 13.2.

In other cases, the knowledge of the viscous properties must also be known if the stresses are to be determined and this will require further equations (see Chapter 8). These equations were independently derived by Navier (1822) and Stokes (1842). Stokes used these equations to solve non-trivial problems.

9.3 Inertial set

It needs to be remembered that while deriving the equations of motion, an assumption was made that Newton's laws of motion are valid in the coordinate system that was used. These laws are, of course, valid only in **inertial** systems as it was recognized already by Galileo Galilee in the 17th century. He stated that rest and motion are two equivalent states of mobility that can be distinguished only in relation to a body in its vicinity.

A reference system linked to the surface of Earth can be considered to be inertial. Any other reference system that is at rest (the laboratory) with regard to it or in a steady linear motion (the observation of actions taking place in a vehicle by a person in this vehicle) is also inertial. In the case of a curvilinear motion or a speed change, additional apparent outside forces would have to be added to account for the change of velocity and direction of the non-inertial system. A centrifugal and the Coriolis force would have to be added for uniform rotary motion.

A simple laboratory experiment can demonstrate that neither the laboratory nor the Earth is the relevant inertial systems. It is the whole Solar System.

The rotation of the Earth can be observed in the laboratory from the motion of a pendulum. The pendulum preserves the plane of the swing in relation to the starry sky (Foucault's pendulum) so that it rotates by 360 degree in a day and once around in a year.

Due to the rotation of the Earth, all currents flowing from the pole to the equator fall behind the rotation of the Earth toward the West. On the other hand, all currents flowing from the equator to the pole get accelerated toward the East. This phenomenon affects not only the winds and ocean currents but it can be also observed when draining liquid from a container (e.g., the washbasin). People living in the northern hemisphere will see a vortex turning anticlockwise whereas people living on the southern hemisphere will see the vortex turning clockwise.

9.4 The momentum balance in a fluid having a constant density and viscosity: Navier-Stokes (N-S) equation

Although the tensor $\boldsymbol{\tau}$ may depend on the velocity distribution in a fluid by a relatively complex relationship, the contribution of friction to the variation of momentum is, fortunately, for the incompressible ($\rho=$const.), and Newtonian ($\mu=$const.) fluids are substantially simpler. It has the form of a **viscous term**

$$\nabla \cdot \boldsymbol{\tau} = \mu\left(\nabla^2 \boldsymbol{u} + \nabla \cdot \boldsymbol{u}\nabla\right) \tag{9.15}$$

The second term in the bracket is, according to the equation of continuity, equal to zero for all liquids having a constant density so that

$$\nabla \cdot \boldsymbol{\tau} = \mu \nabla^2 \boldsymbol{u} \tag{9.16}$$

For the fluid that has a constant density and viscosity, the equation of motion expressed in stresses the stresses can be replaced by velocities to obtain the **Navier-Stokes equation**

$$\rho\left(\frac{\partial \boldsymbol{u}}{\partial t} + \boldsymbol{u} \cdot \nabla \boldsymbol{u}\right) = -\nabla P + \mu \nabla^2 \boldsymbol{u} \tag{9.17}$$

This equation has only four independent variables: the scalar P and three components of the vector $\boldsymbol{u}$. The vector Navier-Stokes equation represents that three equations, with the continuity equation and with the relevant boundary and initial conditions are, in principle, solvable. These are, however, nonlinear equations so that under certain circumstances (high Reynolds numbers—see later), the set has multiple solutions. The flow behaves accordingly—it results in turbulence.

9.5 Summary

Verbally, the general balance in a moving medium can be expressed as

the inflow = the outflow + accumulation

In a moving medium, a balance of miscellaneous physical variables can be performed in a small immobile differential element.

The mass balance is equivalent to the continuity equation that has a character of a differential equation. This equation has a particularly simple form for a material having a constant density in which case the accumulation is zero.

In a momentum balance, the accumulation is related to the forces acting on an element. Usually, the volume forces are related to the action of gravity. Plane forces are represented by the mechanical stresses composed of the tensors of pressure and that of dynamic stress. Usually, a potential of outside forces is introduced that summarizes the influence of pressure and of the volume forces. The mathematical expression of the momentum balance in terms of stresses is the Cauchy equation.

As a rule, the momentum balance is derived for inertial systems, that is for such a system that are related to some reference subject (such as the Earth or, even better, to a Solar System) that is either at rest or moving in a straightforward, uniform motion. Because the distribution of dynamic stresses is not trivial, they must be expressed being dependent on the kinematic tensor. In the simplest case of an incompressible (having an essentially constant density) fluid that displays a rheological behavior of a Newtonian liquid, the Cauchy equation becomes the Navier-Stokes equation.

PART TWO

Balances in transport phenomena

CHAPTER TEN

Solutions of the Navier-Stokes equation

Contents

In the preceding chapter, the momentum balance in a fluid that has a constant density and viscosity was termed the **Navier-Stokes equation**

$$\rho\left(\frac{\partial \boldsymbol{u}}{\partial t}+\boldsymbol{u}\cdot\nabla\boldsymbol{u}\right)=\rho\boldsymbol{g}-\nabla p+\mu\nabla^2\boldsymbol{u} \tag{10.1}$$

With the use of the potential $\nabla P \equiv \nabla p - \rho \boldsymbol{g}$, Eq. (10.1) becomes

$$\rho\left(\frac{\partial \boldsymbol{u}}{\partial t}+\boldsymbol{u}.\nabla\boldsymbol{u}\right)=-\nabla P+\mu\nabla^2\boldsymbol{u} \tag{10.2}$$

If the density of the fluid changes due to temperature or pressure (in the case of gases) variation or if the viscosity changes due to temperature variation, these changes are not so precipitous to invalidate the local validity of the Navier-Stokes equation. However, this equation needs to be modified for non-Newtonian liquids.

In principle, these equations are valid also for the turbulent flow but they do not offer any solution for the generation and for the motion of the turbulent vortices. They

Transport and Surface Phenomena
https://doi.org/10.1016/B978-0-12-818994-8.00010-5

will also fail for extremely fast flows, as they are the sonic and supersonic motions. Additional thermodynamic laws would have to be included to obtain correct balances.

Because Eq. (10.1) is a vector equation, it has the capability of three equations. Along with the continuity equation and the applicable boundary and initial conditions, it has the capability to fully describe the distribution of all three components of the velocity vector and of the scalar quantity of pressure in space and time.

10.1 Cartesian coordinate set

By splitting the Navier-Stokes equation into its components in the **Cartesian coordinate system** (x, y, z), one gets

For the x direction

$$\rho\left(\frac{\partial u_x}{\partial t}+u_x\frac{\partial u_x}{\partial x}+u_y\frac{\partial u_x}{\partial y}+u_z\frac{\partial u_x}{\partial z}\right)=\rho g_x-\frac{\partial p}{\partial x}+\mu\left(\frac{\partial^2 u_x}{\partial x^2}+\frac{\partial^2 u_x}{\partial y^2}+\frac{\partial^2 u_x}{\partial z^2}\right) \tag{10.3}$$

For the y direction

$$\rho\left(\frac{\partial u_y}{\partial t}+u_x\frac{\partial u_y}{\partial x}+u_y\frac{\partial u_y}{\partial y}+u_z\frac{\partial u_y}{\partial z}\right)=\rho g_y-\frac{\partial p}{\partial y}+\mu\left(\frac{\partial^2 u_y}{\partial x^2}+\frac{\partial^2 u_y}{\partial y^2}+\frac{\partial^2 u_y}{\partial z^2}\right) \tag{10.4}$$

and for the z direction

$$\rho\left(\frac{\partial u_z}{\partial t}+u_x\frac{\partial u_z}{\partial x}+u_y\frac{\partial u_z}{\partial y}+u_z\frac{\partial u_z}{\partial z}\right)=\rho g_z-\frac{\partial p}{\partial z}+\mu\left(\frac{\partial^2 u_z}{\partial x^2}+\frac{\partial^2 u_z}{\partial y^2}+\frac{\partial^2 u_z}{\partial z^2}\right) \tag{10.5}$$

Using the vector operator, the Navier-Stokes equation can also be expressed in other coordinate systems.

10.2 Cylindrical coordinate set

Specifically, the Navier-Stokes equations in the **cylindrical coordinate system** (r, z, φ) will take the form

For the radial direction r

$$\rho\left(\frac{\partial u_r}{\partial t}+u_r\frac{\partial u_r}{\partial r}+\frac{u_\varphi}{r}\frac{\partial u_r}{\partial \varphi}-\frac{u_\varphi^2}{r}+u_z\frac{\partial u_r}{\partial z}\right)$$
$$=\rho g_r-\frac{\partial p}{\partial r}+\mu\left[\frac{\partial}{\partial r}\left(\frac{1}{r}\frac{\partial}{\partial r}(ru_r)\right)+\frac{1}{r^2}\frac{\partial^2 u_r}{\partial \varphi^2}-\frac{2}{r^2}\frac{\partial u_\varphi}{\partial \varphi}+\frac{\partial^2 u_r}{\partial z^2}\right] \quad (10.6)$$

For the angular direction φ

$$\rho\left(\frac{\partial u_\varphi}{\partial t}+u_r\frac{\partial u_\varphi}{\partial r}+\frac{u_\varphi}{r}\frac{\partial u_\varphi}{\partial \varphi}+\frac{u_r u_\varphi}{r}+u_z\frac{\partial u_\varphi}{\partial z}\right)$$
$$=\rho g_\varphi-\frac{1}{r}\frac{\partial p}{\partial \varphi}+\mu\left[\frac{\partial}{\partial r}\left(\frac{1}{r}\frac{\partial}{\partial r}(ru_\varphi)\right)+\frac{1}{r^2}\frac{\partial^2 u_\varphi}{\partial \varphi^2}-\frac{2}{r^2}\frac{\partial u_r}{\partial \varphi}+\frac{\partial^2 u_\varphi}{\partial z^2}\right] \quad (10.7)$$

While for the axial direction z

$$\rho\left(\frac{\partial u_z}{\partial t}+u_r\frac{\partial u_z}{\partial r}+\frac{u_\varphi}{r}\frac{\partial u_z}{\partial \varphi}+u_z\frac{\partial u_z}{\partial z}\right)$$
$$=\rho g_z-\frac{\partial p}{\partial z}+\mu\left[\frac{1}{r}\frac{\partial}{\partial r}\left(r\frac{\partial u_z}{\partial r}\right)+\frac{1}{r^2}\frac{\partial^2 u_z}{\partial \varphi^2}+\frac{\partial^2 u_r}{\partial z^2}\right] \quad (10.8)$$

10.3 Spherical coordinate set

In the **spherical coordinate system** (s, φ, θ), the Navier-Stokes equation assumes the following form: for the spherical distance s from the center:

$$\rho\left(\frac{\partial u_s}{\partial t}+u_s\frac{\partial u_s}{\partial s}+\frac{u_\theta}{s}\frac{\partial u_s}{\partial \theta}+\frac{u_\varphi}{s\sin\theta}\frac{\partial u_s}{\partial \varphi}-\frac{u_\theta^2+u_\varphi^2}{s}\right)$$
$$=\rho g_s-\frac{\partial p}{\partial s}+\mu\left[\frac{1}{s^2}\frac{\partial}{\partial s}\left(\frac{\partial}{\partial s}(s^2u_s)\right)+\frac{1}{s^2\sin\theta}\frac{\partial}{\partial \theta}\left(\sin\theta\frac{\partial u_s}{\partial \theta}\right)\right.$$
$$\left.+\frac{1}{s^2\sin\theta}\frac{\partial^2 u_s}{\partial \varphi^2}-\frac{2}{s^2}u_s-\frac{2}{s^2}\frac{\partial u_\theta}{\partial \theta}-\frac{2}{s^2}u_\theta\,\mathrm{cotg}\,\theta\right] \quad (10.9)$$

For the angular direction φ

$$\rho\left(\frac{\partial u_\varphi}{\partial t}+u_s\frac{\partial u_\varphi}{\partial s}+\frac{u_\theta}{s}\frac{\partial u_\varphi}{\partial \theta}+\frac{u_\varphi}{s\sin\theta}\frac{\partial u_\varphi}{\partial \varphi}+\frac{u_s u_\theta}{s}-\frac{u_\varphi u_\theta \operatorname{cotg}\theta}{s}\right)$$
$$=\rho g_\varphi-\frac{1}{s\sin\theta}\frac{\partial p}{\partial \varphi}+\mu\left[\frac{1}{s^2}\frac{\partial}{\partial s}\left(\frac{\partial}{\partial s}\left(s^2 u_\varphi\right)\right)+\frac{1}{s^2\sin\theta}\frac{\partial}{\partial \theta}\left(\sin\theta\frac{\partial u_\varphi}{\partial \theta}\right)\right.$$
$$\left.+\frac{1}{r^2\sin\theta}\frac{\partial^2 u_\varphi}{\partial \varphi^2}-\frac{u_\varphi}{s^2\sin^2\theta}+\frac{2}{s^2\sin^2\theta}\frac{\partial u_s}{\partial \varphi}+\frac{2\cos\theta}{s^2\sin^2\theta}\frac{\partial u_\theta}{\partial \varphi}\right] \tag{10.10}$$

and for the direction of the deflection θ from the axis z

$$\rho\left(\frac{\partial u_\theta}{\partial t}+u_s\frac{\partial u_\theta}{\partial s}+\frac{u_\theta}{s}\frac{\partial u_\theta}{\partial \theta}+\frac{u_\varphi}{s\sin\theta}\frac{\partial u_\theta}{\partial \varphi}+\frac{u_s u_\theta}{s}-\frac{u_\varphi^2 \operatorname{cotg}\theta}{s}\right)$$
$$=\rho g_s-\frac{1}{s}\frac{\partial p}{\partial \theta}+\mu\left[\frac{1}{s^2}\frac{\partial}{\partial s}\left(\frac{\partial}{\partial s}\left(s^2 u_\theta\right)\right)+\frac{1}{s^2\sin\theta}\frac{\partial}{\partial \theta}\left(\sin\theta\frac{\partial u_\theta}{\partial \theta}\right)\right.$$
$$\left.+\frac{1}{s^2\sin\theta}\frac{\partial^2 u_\theta}{\partial \varphi^2}+\frac{2}{s^2}\frac{\partial u_\theta}{\partial \theta}-\frac{u_\theta}{s^2\sin^2\theta}-\frac{2\cos\theta}{s^2\sin^2\theta}\frac{\partial u_\varphi}{\partial \varphi}\right] \tag{10.11}$$

In no case are these equations analytically solvable in the whole range of the variables. However, in some selected simplified cases for which the symmetry of the boundary and initial conditions applies, a number of terms might be zero. Such a circumstance may occur when a velocity component is zero or pressure is zero or it is derivative along the chosen variable is zero. This procedure will lead reliably to simplified equations for momentum balance. To some extent, this procedure may be even more comfortable than the individual balances as it was shown for the viscometric flows.

10.4 Non-inertial rotating set

In the special case of an **axially symmetric non-inertial set rotating with a uniform angular speed around the axis z**, it is customary to formulate the equations as being seen by an observer rotating with the set. In such a case, it is more convenient to replace the peripheral velocity by an angular velocity

$$\omega \equiv u_\varphi / r \tag{10.12}$$

In a **rotating cylindrical set** (r, z, φ), the Navier-Stokes equation becomes, assuming an axial symmetry $\frac{\partial}{\partial \varphi}=0$

For the radial direction r

$$\rho\left(\frac{\partial u_r}{\partial t}+u_r\frac{\partial u_r}{\partial r}+u_z\frac{\partial u_r}{\partial z}\right)=\rho g_r-\frac{\partial p}{\partial r}+\varrho\omega^2 r$$
$$+\mu\left[\frac{\partial}{\partial r}\left(\frac{1}{r}\frac{\partial}{\partial r}(ru_r)\right)+\frac{\partial^2 u_r}{\partial z^2}\right] \qquad (10.13)$$

For the angular direction φ: it will be

$$\rho r\left(\frac{\partial \omega}{\partial t}+u_r\frac{\partial \omega}{\partial r}+u_z\frac{\partial \omega}{\partial z}\right)=\rho g_\varphi-2\rho\omega u_r$$
$$+\mu\left[\frac{\partial}{\partial r}\left(\frac{1}{r}\frac{\partial}{\partial r}(r^2\omega)\right)+r\frac{\partial^2 \omega}{\partial z^2}\right] \qquad (10.14)$$

While for the axial direction z

$$\rho\left(\frac{\partial u_z}{\partial t}+u_r\frac{\partial u_z}{\partial r}+u_z\frac{\partial u_z}{\partial z}\right)=\rho g_z-\frac{\partial p}{\partial z}+\mu\left[\frac{1}{r}\frac{\partial}{\partial r}\left(r\frac{\partial u_z}{\partial r}\right)++\frac{\partial^2 u_r}{\partial z^2}\right] \qquad (10.15)$$

Two terms belonging originally to the inertial forces have been shifted to the right-hand side as they act as additional apparent mass forces, i.e.,
– Centrifugal force $\varrho\omega^2 r$
– Coriolis force $2\rho\omega u_r$
Similarly, the equations in the spherical coordinate system can also be modified.

10.5 The estimation of the importance of the individual terms in the Navier-Stokes equation. The Reynolds number

There are three groups of terms in the equation of motion.

The term on the left-hand side of Eq. (9.17), that is the change of momentum,

$$\rho\left(\frac{\partial \boldsymbol{u}}{\partial t}+\boldsymbol{u}.\nabla\boldsymbol{u}\right)$$

due to the fact that the motion is neither uniform nor straightforward. In other words, it characterizes the change of inertia, and therefore it is called the **inertial term.**

The term on the right-hand side of the Eq. (9.17), $\nabla\boldsymbol{\tau}$ is called the **friction term**

The combination of the mass force and the pressure expressed by the quantity $\nabla P \equiv \rho \mathbf{g} - \nabla p$ is known as the **driving force**.

In the following, it is attempted to evaluate the magnitude of these terms in space positioned in a Cartesian coordinate system, where X is a characteristic length in the direction x and Y is a characteristic length in the direction y. There is no change in the z direction. The speed in the x direction is characterized by a quantity U. The appropriate choices are based on the requirement to have, in the region of interest, the absolute value of the ratio u_x/U move somewhere between zero and a unitary number. This makes it possible to use, for the velocity U, either the maximum value of u_x or some average speed. The symbolic statement

$$u_x = o(U) \tag{10.16}$$

is pronounced: u_x is **on the order of** U.

10.5.1 The detail of the derivation

Assuming that the speed varies continuously it may be expected

$$\frac{\partial u_x}{\partial x} = o\left(\frac{U}{X}\right) \tag{10.17}$$

$$\frac{\partial^2 u_x}{\partial x^2} = o\left(\frac{U}{X^2}\right) \tag{10.18}$$

$$\frac{\partial^2 u_x}{\partial y^2} = o\left(\frac{U}{Y^2}\right) \tag{10.19}$$

On the basis of the equation of continuity, it should hold

$$\frac{\partial u_y}{\partial y} = o\left(\frac{U}{X}\right) \tag{10.20}$$

and therefore also

$$u_y = o\left(\frac{U\,Y}{X}\right) \tag{10.21}$$

etc.

If the set keeps changing for a characteristic time span t^*, then it will also hold

$$\frac{\partial u_x}{\partial t} = o\left(\frac{U}{t^*}\right) \tag{10.22}$$

10.5.2 Result

The formalism used was introduced in order to assess the magnitude of the individual terms of the Navier-Stokes equation. The ordinal dimension of the inertial term in the x direction is

$$\rho\left(\frac{\partial u_x}{\partial t}+u_x\frac{\partial u_x}{\partial x}+u_y\frac{\partial u_x}{\partial y}\right)=o\left[\frac{\rho\,U^2}{X}\left(\frac{U}{X\,t^*}+1\right)\right] \tag{10.23}$$

If the characteristic time t^* is large enough to meet the condition

$$\frac{U}{X\,t^*}\ll 1 \tag{10.24}$$

then the time term in the equation can be neglected and the process could be considered as steady.

The dimension of the friction term is

$$\mu\left(\frac{\partial^2 u_x}{\partial x^2}+\frac{\partial^2 u_x}{\partial y^2}\right)=o\left[\frac{\mu\,U}{Y^2}\left(1+\left(\frac{Y}{X}\right)^2\right)\right] \tag{10.25}$$

The ratio of the inertial term and the friction term can be estimated from the magnitude of the fraction

$$\mathrm{Re}\equiv\frac{\rho\,U\,Y}{\mu} \tag{10.26}$$

which is habitually known as the **Reynolds number** in honor of the scientist who, in 1883, found experimentally that the **creeping flow** in tubes of the circular cross section is maintained as long as *Re* is smaller than about 2000. The variables in the expression (10.26) are U is the average velocity, Y is the cross section of the tube, and ρ is the density of the liquid. This is remarkable because for steady flows in straight tubes of any cross section (not only circular) is the term $\boldsymbol{u}\cdot\nabla\boldsymbol{u}$ identically equal to zero so that the inertial forces should not play any role. Accidental disturbances causing a deviation from a straightforward flow (such as an insignificant roughness) will be dampened at low Reynolds numbers, while they might become magnified at high *Re*. This effect might lead to a jump change that might lead to turbulence.

Flows with curved streamlines show the influence of inertial forces already at Reynolds numbers in the range between one and ten. Initially, they bring about a uniform whirling that, at higher Reynolds number, continuously changes to chaotic turbulence. It needs to be emphasized, however, that although chaos in mathematical and physical terms means a considerable unpredictability it does follow certain stochastic regularity (=obeys the rules of statistics).

10.6 The effect of the Reynolds number

10.6.1 Creeping flow

The Navier-Stokes equation can be simplified depending on the value of the Reynolds number. The inertial term can be neglected for flows at low Reynolds numbers so that the Navier-Stokes equation becomes the **creeping flow equation**

$$\frac{\partial \boldsymbol{u}}{\partial t} = \frac{-\nabla P}{\rho} + \frac{\mu}{\rho}\nabla^2 \boldsymbol{u} \text{ for Re} \ll 1 \tag{10.27}$$

This equation has a linear mathematical structure, similar to the equation of heat conduction or the equation of diffusion with a source term so that it offers the same class of easily solvable tasks. Software programs are available that make it possible for a programmer to assign the appropriate boundary conditions. The computer will then find the solution without further specification and present it in the chosen format. The memory requirement and the time needed for finding the solution depend on the complexity of the task and on the required accuracy. A particularly simple is the solution for a steady flow for which the accumulation term is equal to zero.

10.6.2 The inviscid flows

Conversely, if the Reynolds number is too large then the friction term in the Navier-Stokes equation can be ignored and the equation reduces to the **Euler equation for the flow of an ideal fluid**

$$\frac{\partial \boldsymbol{u}}{\partial t} + \boldsymbol{u} \cdot \nabla \boldsymbol{u} = \frac{-\nabla P}{\rho} \text{ for Re} \ll 1 \tag{10.28}$$

There are, however, even in the case of inviscid flows, regions where the viscosity might become the controlling factor. This region is known as the **boundary layer** (see Section 16.3 and further) arises in the vicinity of large velocity gradients and close to the walls.

The Euler equation is a nonlinear differential equation the solution of which is, at higher Reynolds numbers, difficult to arrive at due to mathematical problems. Also, the uniqueness of any such solution is problematic.

This is not a shortcoming of the mathematical assignment or of the solution. Even in real situations, the system becomes unstable, the flow becomes chaotic and the state is called the **turbulence.**

10.7 Characteristic hydrodynamic quantities

As a rule, the solution of the Navier-Stokes equation takes the form of a **velocity and pressure field,** thus two distributed variables. In the engineering practice, however, such a detailed description of the situation is not required. Instead, the knowledge of the **lumped variables,** such as the characteristic **local or integral quantities** is required. In hydrodynamics, these quantities are usually one of the following: the maximum or the minimum velocity and their location, the volumetric flow, and the average velocity, the values of pressure in selected positions, the overall pressure difference, forces acting on the walls, etc.

The quantitative distribution of the tangential stress can be often estimated from just one measurable **dynamic quantity,** such as for example the pressure difference or the torque moment. The magnitude of local velocities can be obtained from a measurable **kinematic quantity,** such as the speed of rotation of a wall or from the flow rate of the liquid. If such a pair is known, then in the case of a creeping flow, the **dynamic viscosity** can be determined from measured values with the use of an uncomplicated mathematical theory. The term **viscometric flows**, used for simple steady creeping flows, was derived from this setup. In the reverse case, from a known viscosity one of the pairs mentioned above can be calculated

10.8 The boundary and initial conditions

To solve a transport problem, it is useful to find the simplest system bounded by space and time for which the **initial conditions** are known and for which the distribution of the examined property at the boundary, **boundary conditions**, is also known. Particularly important are the **steady transports** for which the boundary conditions last such a long time so that the final state of the system essentially does not depend on the initial state.

The question about how many **boundary and initial conditions** are needed to solve unambiguously a partial differential equation can be easily answered from the kind of derivation that in the equation occur. The first-order derivative by time requires just one initial condition, the knowledge of the distribution of the dependent variable (e.g., the speed and the pressure) at a specific time. The existence of the second-order derivative by location at the point $\boldsymbol{x}$ in the Cartesian coordinate system means that two pieces of information are needed about the distribution of the dependent variable depending on $\boldsymbol{x}$ at any time and for all values of $\boldsymbol{y}$ and $\boldsymbol{z}$ comprising the studied task. By analogy, two pieces of information are needed about the dependence on $\boldsymbol{y}$ and $\boldsymbol{z}$. If in the given direction only the first derivative exists then only one value will suffice. Naturally, for the trivial case when the process does not depend on any of the variables (x, y, z, or t), the particular derivative will fall out from the equation of heat conduction so that the conditions linked to it will not be needed.

The more it becomes possible to express the real problem by a system geometrically simpler, the simpler will be the mathematical model of the transport process and the more lucid will become the mathematical solution. In addition, the clearer will be the results.

10.9 Summary

In the momentum balance of a unitary volume that is in the Navier-Stokes equation expressed in vectors, the first term on the left is the accumulation of momentum while the second is the inflow and the outflow of momentum. The left-hand side represents the inertial terms. On the right side is the sum of forces that act on the element. Here, apart from the gravity is the surface force split between the pressure and the friction. In some cases, it is advantageous to sum up the driving forces, which are the pressure and the gravity in a single potential of outside forces.

The vector form of the Navier-Stokes equation can be broken down into three scalar equations in one of the spatial coordinate systems. Such a step is useful if, due to the symmetry some terms in these equations are identically equal to zero.

In the case of a uniform rotation with regard to these immobile objects, the influence of such rotation on the transport of momentum, from the point of view of an observer rotating with the object, can be expressed with two apparent mass forces; the centrifugal and the Coriolis forces.

A specific solution of the Navier-Stokes equation requires knowledge of two boundary conditions on the border of the system and one initial condition.

If boundary and initial conditions are given then it is possible to make a rough estimate of the magnitude of the individual terms of the Navier-Stokes equation by making a suitable choice of the characteristic length and the characteristic velocity. This choice will lead to the definition of the Reynolds number which estimates, for a greater part of the studied problem, the numerical share of the importance of the ratio of the inertial to the friction terms.

The flow characterized by a low Reynolds number that is by a very high viscous term is known as the creeping flow. The distribution of velocities and forces can be very well approximated by solving the equation of the creeping flow that is the Navier-Stokes equation in which the term of expressing the inflow and outflow of momentum is omitted (i.e., the second, eventually the first term on the left-hand side).

The flow at very high Reynolds numbers, thus the flow where the inertial forces play a controlling role, can be studied, under certain conditions, with the help of the equation of the inviscid flow described by the Navier-Stokes equation in which the friction term (the last term on the right-hand side) is left out. In effect, even in this case, there is a small region, usually at the wall, where the viscous friction is significant. Outside this boundary layer, the flow is chaotic. It is known as turbulent.

CHAPTER ELEVEN

The balance of mass and of mechanical energy in a flow tube

Contents

11.1 The integral equation of continuity and the Bernoulli equation

It is possible to gain some information about the flow of fluids by solving the balance of mechanical energy. The pioneer of this method was Daniel Bernoulli. The balance relates to a **flow tube**, which can be a conduit bounded by just imagined surfaces formed by streamlines or a tube bounded by real solid walls that has an inner volume V with one reference plane S_1 through which a fluid of density ρ_1 enters at a speed u_1 and with other reference exit plane S_2 through which a fluid of density ρ_2 leaves at a speed u_2 (see Fig. 11.1).

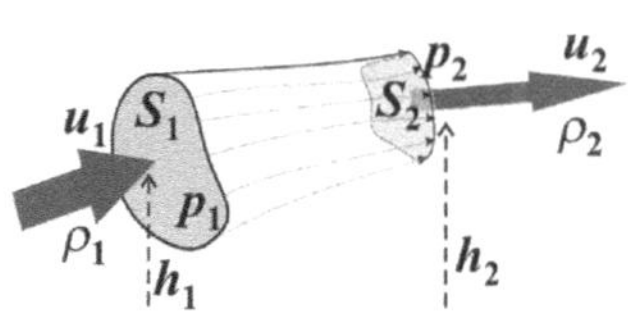

Fig. 11.1 A flow tube between the reference planes S_1 and S_2.

Also, in this case, the mass balance is represented by the **continuity equation** that can be written as

$$V\frac{d\rho_m}{dt} = \rho_1 \mathbf{S}_1 \cdot \mathbf{u}_1 - \rho_2 \mathbf{S}_2 \cdot \mathbf{u}_2 \tag{11.1}$$

https://doi.org/10.1016/B978-0-12-818994-8.00011-7

in which the index m denotes the mean value of the quantity in volume V. If the density is constant, then relation (11.1) becomes

$$\boldsymbol{u} \cdot \boldsymbol{S} = const. \tag{11.2}$$

In the balance of **mechanical energy in a unit of volume** E/V, two terms stand out, which have already been dealt with in the mechanics of a mass point:

1. **the potential energy** is, at the entry point, equal to $\rho_1 h_1 g$, where h_1 is the geodetic height above the chosen zero potential levels in the gravitation field.
2. **the kinetic energy** is, at the entry point, equal to $\rho_1 u_1^2/2$.

In addition, there is also the volume energy that is dealt with in thermodynamics. This discipline deals only with systems in which both the kinetic and the potential energy do not change but the pressure energy is the only component which shows the variation in the system and is determined by a formula $\int p dV$. Thus, the pressure energy in the element of the volume is equal to the pressure.

3. **the pressure energy** is, at the entry point, equal to the pressure p_1.

Thus, in summary, the time change dE/dt of the mechanical energy in the flow tube is given by the difference between the inflow and outflow

$$\frac{dE}{dt} = u_1 S_1 \left(\rho_1 g h_1 + \frac{\rho_1 u_1^2}{2} + p_1 \right) - u_2 S_2 \left(\rho_2 g h_2 + \frac{\rho_2 u_2^2}{2} + p_2 \right) \tag{11.3}$$

This change can be realized in a number of ways, including their combination.

Among the **unsteady flows**, it is worthwhile to mention the start-up of a liquid in a pipeline in which **kinetic energy is accumulated**.

Example: A horizontal pipeline 50 km long that has a diameter of 400 mm is full of incompressible crude oil at rest. The density of the oil is $\rho = 950\,kg/m^3$. A pump is started, which generates virtually a constant pressure $p_1 - p_2 = 2\,MPa$. The power input in the volume of the pipeline is almost entirely due to acceleration

$$\frac{dE}{dt} = \frac{\rho V}{2} \frac{d(u^2)}{dt}$$

How long will it take to achieve the speed of 4 m/s in the pipeline assuming that the power losses can be neglected (95 s).

Another example of unsteady flow is the flow in a tube with flexible walls in which not only the speed but also the **volume of the flow tube** V can change.

An example is the pulsation of the flexible walls of blood vessels.

In the case of **steady flows** it can be assumed, based on the equation of continuity (11.2), a constant mass flow rate

$$\frac{dm}{dt} \equiv \rho\, \boldsymbol{u} \cdot \boldsymbol{S} = const. \tag{11.4}$$

If there are no changes of mass in the system under consideration, that is if the equation of continuity holds, then the change of energy E related to a unit of mass m of the fluid passed is

$$\varepsilon \equiv \frac{dE}{dt} / \frac{dm}{dt} = \left(g h_1 + \frac{u_1^2}{2} + \frac{p_1}{\rho_1} \right) - \left(g h_2 + \frac{u_2^2}{2} + \frac{p_2}{\rho_2} \right) \tag{11.5}$$

The value of ε is positive if the energy is supplied to the system, most frequently by a pump included in the flow tube under consideration. In other cases, the ε is negative when the fluid flowing through the flow tube performs work on its surroundings; either a useful work (e.g., driving a turbine) or its kinetic energy is lost by friction (it dissipates in heat). The conversion of other energies (heat, chemical, electromechanic) to mechanical energy could be either negative or positive.

If the accumulation of energy in the section of the examined flow tube is zero, then Eq. (11.3) reduces in the **Bernoulli** equation

$$\rho g h + \frac{\rho u^2}{2} + p = const. \tag{11.6}$$

11.2 The losses of mechanical energy by friction

Some portion of the energy in a unit of volume is the **energy dissipated** by friction $-\varepsilon_{dis}$, that is, converted in heat. This portion of the energy is called **lost** because in fluids it is not large enough to bring an appreciable increase in temperature.

Example: If the pressure in water flowing through a pipe drops by 0.2 MPa, then the energy lost by friction is $\varepsilon_{dis} = \Delta p = 200\,J/m^3$*, which corresponds to a temperature increase of* $-\varepsilon_{dis}/(\rho \cdot c_P)$ *which is less than 0.1 K. In solid materials, the energy by friction can be concentrated in a small volume of mass (e.g., in automobile breaks or when making fire by friction). Here the heating is significant.*

Heating by dissipative mechanical energy is utilized in the process of kneading polymers to become highly viscous liquids.

Since the loss of energy by friction can occur only during motion, its value is measured by the characteristic magnitude of the kinetic term of the energy balance. Thus, a **coefficient of energy loss** will be defined by

$$\zeta \equiv \Delta(\rho\varepsilon)/\frac{\rho u_c^2}{2} \tag{11.7}$$

where the characteristic speed u is chosen according to whatever is customary in a given task. For example, the mean velocity in a free cross section of a pipe is chosen for local losses in a pipeline.

The Bernoulli equation containing the losses of the mechanical energy can be written as

$$\rho g(h_2 - h_1) + \frac{\rho\left(u_2^2 - u_1^2 + \zeta\, u_c^2\right)}{2} + (p_2 - p_1) = 0 \tag{11.8}$$

The coefficient of energy loss for high Reynolds numbers that are common for most flows of water and air and similar fluids is practically constant.

$$\zeta \to \zeta_\infty \text{ for } \frac{L_c u_c \rho}{\mu} \to \infty \tag{11.9}$$

The flows at low Reynolds numbers that occur either with highly viscous liquids or with slow flows through capillaries and porous media, do not depend much on the density. At very low Reynolds numbers, the loss coefficient is inversely proportional to the characteristic Reynolds number. Therefore, it is convenient to introduce a new quantity

$$Y \equiv \zeta \frac{L_c u_c \rho}{\mu} \tag{11.10}$$

whereby

$$Y \equiv Y_0 \text{ for } \frac{L_c u_c \rho}{\mu} \to 0 \tag{11.11}$$

so that the Bernoulli equation inclusive of losses of the mechanical energy can be written as

$$\rho g(h_2 - h_1) + \frac{\rho\left(u_2^2 - u_1^2\right)}{2} + (p_2 - p_1) + \frac{Y\,\mu\,u_c}{L_c} = 0 \tag{11.12}$$

which is convenient for creeping (laminar) flows when the value $Y = Y_0$ can be obtained by solving the Navier-Stokes equation (see Section 12.1).

11.3 The basic equations of hydrostatics

For a liquid at rest, the Bernoulli equation reduces to the **Pascal law**, the fundamental equation of hydrostatics

$$\rho g h + p = const. \tag{11.13}$$

This equation is used for the interpretation of the pressure differences measured by a liquid pressure gauge and, in a modified form when introducing the power potential (9.13).

In those fluids that change their density according to the circumstances (gases change their density with pressure, temperature, and composition), it is necessary to use an integral relationship for the determination of the dependence of the pressure on the position

$$p_2 = p_1 + \int_{h_1}^{h_2} \rho(h) g \, dh \tag{11.14}$$

This relationship can be used for the determination of the dependence of the atmospheric pressure on the geodetic height when using the density determined by the equation of state of an ideal gas

$$\rho = \frac{Mp}{RT}$$

then the dependence of pressure on the height will be obtained in the form of an exponential relationship that can be linearized for small pressure differences (for a pressure difference of 1000 m the error amounts to 1% and it rises further steeply). In meteorology, the formulas are not only recalculated to the sea level, but they also respect changes in temperature and humidity of the air. Conversely, it is also necessary to apply the recalculations in the altimeters that measure the local pressure. The formulas are set according to the instant meteorological situation.

A special case is the energy balance for a fluid that does not move relative to a uniformly rotating set. For such a case, it is possible to derive from Eq. (10.13) a relationship

$$\rho\left(gh + \frac{\omega r^2}{2}\right) + p = const. \tag{11.15}$$

where r is the distance from the axis of rotation.

11.4 Summary

The basic components of mechanical energy in solid bodies are the kinetic and potential energies. Additionally, pressure energy is the third component in fluids.

The balance of mechanical energy in fluid moving in a real or in an imaginary flow tube, the walls which serve as the boundaries between the reference areas (streamlines) and are impervious to any fluid, can be simplified by using the mass balance (the equation of continuity).

The balance of mechanical energy will be reduced to the Bernoulli equation if there is no transformation of energy (supply of energy or taking it away, or losses due to dissipation to heat).

If, however, energy losses during the flow do occur it is useful to introduce the coefficient of energy loss and to include the loss term in the Bernoulli equation.

For high Reynolds number flows, the coefficients of energy losses are constant, and their values are tabulated. For flows at low Reynolds numbers, the coefficients are inversely proportional to the Reynolds number and its magnitude can be obtained by solving the Navier-Stokes equation.

For fluids at rest, the Bernoulli equation reduces to the Pascal law expressing the dependence of pressure in the fluid on the geodetic height.

CHAPTER TWELVE

The steady unidirectional flow

Contents

12.1 The steady one-dimensional unidirectional creeping flow

12.1.1 The flow through a tube of a constant cross section

A flow with a high degree of symmetry is a setup in which, in a given coordinate system, there is only one nonzero component of the velocity vector (Fig. 12.1).

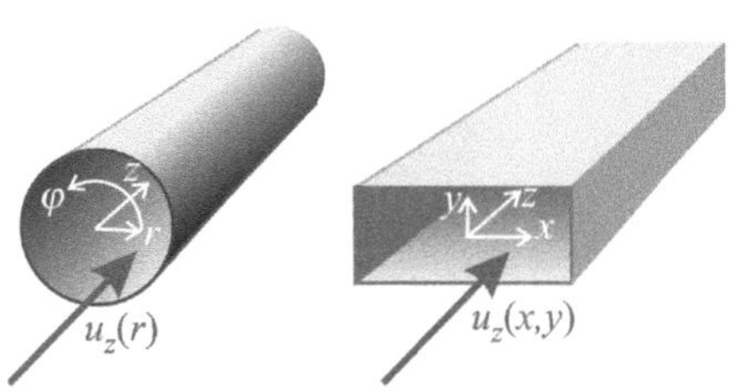

Fig. 12.1 Unidirectional (*z*) longitudinal flow through a tube of a constant cross section. One-dimensional (*r*); two-dimensional (*x*, *y*).

Transport and Surface Phenomena
https://doi.org/10.1016/B978-0-12-818994-8.00012-9

An example of the one-dimensional problem in such a setup, where the velocity depends exclusively on one spatial coordinate, is just the laminar flow in a tube of circular cross section (Fig. 7.1). The intuitive solution of that problem was already dealt with in Section 9.1.

Starting with a Navier-Stokes equation in cylindrical coordinates from which all zero velocities ($u_r = 0$, $u_\varphi = 0$) as well as all irrelevant derivation $\partial/\partial t = 0$, $\partial/\partial\varphi = 0$, $\partial/\partial z = 0$, will be removed then, assuming that

$$\rho g_z - \frac{\partial p}{\partial x} \equiv -\frac{\partial P}{\partial x} = const. \equiv -\frac{\Delta P}{L} \tag{12.1}$$

an ordinary differential equation is obtained

$$0 = -\frac{\Delta P}{L} + \mu \frac{1}{r}\frac{d}{dr}\left(r\frac{du_z}{dr}\right) \tag{12.2}$$

the solution of which, with relevant boundary conditions, is the velocity profile $u_z(r)$ (see Eq. 7.5) and its integration results in the Hagen-Poiseuille law (7.6). It can be also formulated in a dimensionless form

$$\frac{\Delta P d^4}{\mu L V'} = \frac{128}{\pi} \tag{12.3}$$

or

$$\frac{\Delta P d^2}{\mu L U} = 32 \tag{12.4}$$

Frequently, this group is reformulated after adding another variable, the density, to yield the Reynolds number

$$Re \equiv \frac{U d \rho}{\mu} \tag{12.5}$$

and the **coefficient of friction loss** that characterizes the ratio of the pressure change to the kinetic energy of a unit volume

$$\Lambda \equiv \frac{\Delta P}{\frac{\rho}{2} U^2} \frac{d}{L} \tag{12.6}$$

so that relation (12.4) acquires the common form

$$\Lambda = \frac{64}{Re} \text{ (for } Re < 2000\text{)} \tag{12.7}$$

The forces on a wall are characterized by the tangential stress that, according to Eq. (7.1), is equal to

$$\tau_W = \frac{\Delta P d}{2L} \tag{12.8}$$

The friction coefficient is sometimes replaced by a dimensionless factor f, generally known as the **Fanning friction factor**,

$$f \equiv \frac{\tau_W}{\frac{\rho}{2} U^2} \tag{12.9}$$

In some engineering literature, particularly in the US, the Fanning friction factor is preferred before the coefficient of friction. One of the reasons for this custom is in the absence of any geometric variable in the Fanning factor so that it can be used in a multitude of situations.

For the laminar flow in a tube of circular cross sections it will be then

$$f = \frac{16}{Re} = \frac{\Lambda}{4} \tag{12.10}$$

The simple definition of the friction factor will be used also in this work for the dimensionless description of the dynamic behavior of fluids in some more complex situations close to the wall. It can be used even in those cases when the tangential stress is not constant.

Similarly, it is possible to solve other unidirectional one-dimensional viscometric flows in a conduit of annular cross sections (Fig. 12.2) where the velocity profile is also $u_z(r)$. In a flat slot the velocity profile is $u_y(x)$.

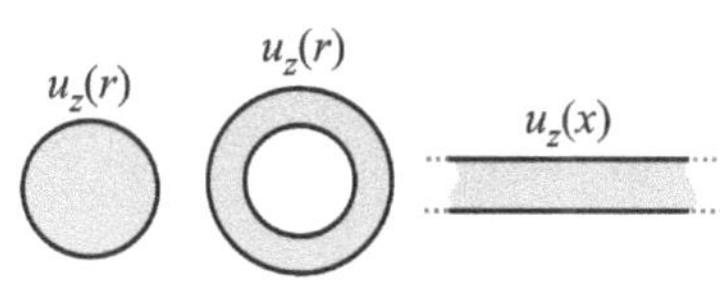

Fig. 12.2 Cross sections of the channels for unidirectional flow.

12.1.2 Unidirectional rotational flows

Apart from the flows mentioned above, there are also unidirectional rotational flows. In particular, this is the rotation between two coaxial cylinders (Section 7.3). Here the sought-after function in cylindrical coordinates is (under the assumption of laminar flow) the only nonzero velocity component $u_\varphi(r)$.

Among the unidirectional rotational flows belongs also the rotation between two parallel round plates for which the sought-after function will be $\omega(z)$, where $\omega \equiv u_\varphi / r$. Similarly, for the flow between two coaxial cones, the sought function in spherical coordinates will be $\omega(\theta)$, see Fig. 12.3.

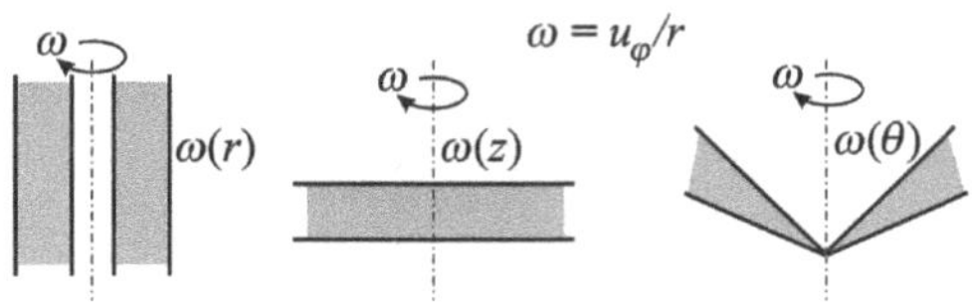

Fig. 12.3 One-dimensional systems of rotational flow. Double cylinder; double plate; double cone.

The flows in these geometries are also used to viscometric measurements as was shown in Fig. 7.4.

12.2 Possible solutions of the equation of flow in terms of stresses

If the assumption of a **steady creeping flow** being an asymptote for the region of very low Reynolds numbers is acceptable for a particular flow situation, then the Cauchy equation in terms of stresses will become

$$0 = -\nabla P + \nabla . \boldsymbol{\tau} \tag{12.11}$$

while the Navier-Stokes equation will get simplified to

$$0 = -\nabla P + \mu \nabla^2 \boldsymbol{u} \tag{12.12}$$

The Cauchy equation (9.14) can be broken down for steady creeping flows in:

Cartesian coordinates:

$$\frac{\partial P}{\partial x} = \frac{\partial \tau_{xx}}{\partial x} + \frac{\partial \tau_{yx}}{\partial y} + \frac{\partial \tau_{zx}}{\partial z} \tag{12.13}$$

$$\frac{\partial P}{\partial y} = \frac{\partial \tau_{xy}}{\partial x} + \frac{\partial \tau_{yy}}{\partial y} + \frac{\partial \tau_{zy}}{\partial z} \tag{12.14}$$

$$\frac{\partial P}{\partial z} = \frac{\partial \tau_{xz}}{\partial x} + \frac{\partial \tau_{yz}}{\partial y} + \frac{\partial \tau_{zz}}{\partial z} \tag{12.15}$$

Cylindrical coordinates:

$$\frac{\partial P}{\partial r} = \frac{1}{r}\frac{\partial (r\,\tau_{rr})}{\partial r} + \frac{1}{r}\frac{\partial \tau_{\varphi r}}{\partial \varphi} - \frac{\tau_{\varphi r}}{r} + \frac{\partial \tau_{zr}}{\partial z} \tag{12.16}$$

$$\frac{1}{r}\frac{\partial P}{\partial \varphi} = \frac{1}{r^2}\frac{\partial \left(r^2\,\tau_{r\varphi}\right)}{\partial r} + \frac{1}{r}\frac{\partial \tau_{\varphi\varphi}}{\partial \varphi} + \frac{\partial \tau_{z\varphi}}{\partial z} \tag{12.17}$$

$$\frac{\partial P}{\partial z} = \frac{1}{r}\frac{\partial (r\,\tau_{rz})}{\partial r} + \frac{\partial \tau_{\varphi z}}{\partial \varphi} + \frac{\partial \tau_{zz}}{\partial z} \tag{12.18}$$

Spherical coordinates:

$$\frac{\partial P}{\partial s} = \frac{1}{s^2}\frac{\partial (s^2\,\tau_{ss})}{\partial s} + \frac{1}{s\sin\theta}\frac{\partial (\tau_{\theta s}\sin\theta)}{\partial \theta} + \frac{1}{s\sin\theta}\frac{\partial \tau_{\varphi s}}{\partial \varphi} - \frac{\tau_{\varphi\varphi} + \tau_{\theta\theta}}{s} \tag{12.19}$$

$$\frac{1}{s}\frac{\partial P}{\partial \theta} = \frac{1}{s^2}\frac{\partial (s^2\,\tau_{s\theta})}{\partial s} + \frac{1}{s\sin\theta}\frac{\partial (\tau_{\theta\theta}\sin\theta)}{\partial \theta} + \frac{1}{s\sin\theta}\frac{\partial \tau_{\varphi\theta}}{\partial \varphi} + \frac{\tau_{s\theta}}{s} - \frac{\tau_{\varphi\varphi}\operatorname{cotg}\theta}{s} \tag{12.20}$$

$$\frac{1}{s\sin\theta}\frac{\partial P}{\partial \varphi} = \frac{1}{s^2}\frac{\partial \left(s^2\,\tau_{s\varphi}\right)}{\partial s} + \frac{1}{s}\frac{\partial \tau_{\theta\varphi}}{\partial \theta} + \frac{1}{s\sin\theta}\frac{\partial \tau_{\varphi\varphi}}{\partial \varphi} + \frac{\tau_{s\varphi}}{r} + \frac{2\,\tau_{\theta\varphi}\operatorname{cotg}\theta}{s} \tag{12.21}$$

If the studied flow system displays a high symmetry (plane or axial), then a number of components of the vectors and tensors in a suitable coordinate set will be zeros so that the solution might lead to a simple ordinary differential equation that will facilitate finding the distribution of the active component of the tangential stress.

It was shown in Section 7.2 of this text that, with simple boundary conditions, it is easy to deduce the necessary equations directly without being forced to start with the simplification of the equations of motion (Cauchy or Navier-Stokes). If, however, it is possible on the basis of the symmetry surmise that the tangential stress in some of the Cauchy equations of motion has only one significant (and with it symmetrical) component, and if the distribution of the potential P is known beforehand, then it is convenient to begin from the integration of this equation. This will be shown in the following cases.

12.3 Solving the problems in stresses

12.3.1 The flow in a flat layer

All what is left from the only nonzero equation (12.13) is, in Cartesian coordinates,

$$\frac{dP}{dx} = \frac{\partial \tau_{yx}}{\partial y} \tag{12.22}$$

and after integration

$$\tau_{yx} = \frac{dP}{dx} y + C \tag{12.23}$$

For a given P it is possible to calculate by integration the spatial distribution of the relevant stress component while it is necessary to utilize the boundary conditions for the concretization of a specific integral.

Otherwise, it is possible to skip the determination of the tangential stress and integrate directly the Navier-Stokes equation

$$\frac{dP}{dx} = \mu \frac{d^2 u_x}{dy^2} \tag{12.24}$$

A good example of the use of Eqs. (12.13)–(12.15) is, for example, the solution of the **flow in a flat channel** or the **flow in a film down the wall.** The usual parabolic velocity profile will be obtained for the flow in the x direction.

$$u_x = \frac{dP}{dx} \frac{y^2}{2\mu} + C_1 y + C_0 \tag{12.25}$$

When using the following boundary conditions

$$u_x = 0 \text{ for } y = -h/2 \tag{12.26}$$

$$u_x = 0 \text{ for } y = h/2 \tag{12.27}$$

the following solution is obtained

$$u_x = \frac{dP}{dx} \frac{1}{2\mu} \left[y^2 - \left(\frac{h}{2} \right)^2 \right] \tag{12.28}$$

The problem is easily solvable even if

$$u_x = U_1 \text{ for } y = y_1 \tag{12.29}$$

$$u_x = U_2 \text{ for } y = y_2 \tag{12.30}$$

For the **flow with a free surface** along the coordinate y_2 it will hold that

$$\tau_{yx} = 0 \text{ and then } \frac{du_x}{dy} = 0 \text{ for } y = y_2 \tag{12.31}$$

This additional condition makes it possible to back calculate the thickness of the layer $|y_1 - y_2|$ between the wall and the surface for a gravitational flow where the gradient of the force potential (according to Eq. 12.13) is equal to

$$\frac{dP}{dx} = -\rho g_x \tag{12.32}$$

12.3.2 Longitudinal flow with axial symmetry

In cylindrical coordinates, all what is left from the only nonzero equation (12.18) is

$$\frac{dP}{dz} = \frac{1}{r}\frac{\partial(r\tau_{rz})}{\partial r} \tag{12.33}$$

and after the integration

$$\tau_{rz} = \frac{r}{2}\frac{dP}{dx} + \frac{C}{r} \tag{12.34}$$

Further, with the use of viscosity

$$\frac{dP}{dz} = \frac{\mu}{r}\frac{d}{dr}\left(r\frac{du_z}{dr}\right) \tag{12.35}$$

By the integration of Eq. (12.35), the spatial velocity distribution can be obtained. Eqs. (12.34) and (12.35) will become useful for the analysis of flows in a circular pipe (Eq. 7.5) and also for a **tube of annular cross section**.

The basic boundary conditions for an annular tube are

$$u_z = 0 \text{ for } r = R_1 \tag{12.36}$$

$$u_z = 0 \text{ for } r = R_2 \tag{12.37}$$

For a given pressure loss the velocity profile is obtained by integration as

$$u_z(r) = \frac{R_2^2}{4\mu}\left(\frac{-dP}{dz}\right)\left\{1-\left(\frac{r}{R_2}\right)^2+\left[1-\left(\frac{R_1}{R_2}\right)^2\right]\frac{\ln(r/R_2)}{\ln(R_1/R_2)}\right\} \tag{12.38}$$

The average velocity is the ratio of the volumetric flow rate and the cross section of the conduit

$$U \equiv \frac{V'}{S} = \frac{\int_{R_1}^{R_2} 2\pi\, r\, u_z(r)\, dr}{\pi(R_2^2 - R_1^2)} \tag{12.39}$$

The average magnitude of the tangential stress on the walls is

$$\overline{\tau_{rz}} = \pi\left(R_2^2 - R_1^2\right)\left(\frac{-dP}{dz}\right)/(2\pi R_1 + 2\pi R_2) \tag{12.40}$$

Since the tangential stress in a circular tube of diameter d is

$$\tau_{rz} = \left(\frac{-dP}{dz}\right)\frac{d}{4} \tag{12.41}$$

It is tempting to define **an equivalent diameter** of a noncircular profile by a relationship

$$d_e \equiv \frac{4 \times \text{cross section}}{\text{perimeter}} \tag{12.42}$$

and specifically for an annulus

$$d_e = 2\,(R_2 - R_1) \tag{12.43}$$

If this equivalent diameter is used in the definition of the Reynolds number according to Eq. (12.5), then the product

$$f\,Re \equiv \frac{8d_e^2}{\mu U}\left(\frac{-dP}{dz}\right) == Y(R_1/R_2) \tag{12.44}$$

is a function of the ratio of the diameters of the walls of the annulus. This is graphically visualized in Fig. 12.4. Evidently, there are asymptotes $Y=64$ for $R_1 \rightarrow 0$ and $Y=96$ for $R_1 \rightarrow R_2$.

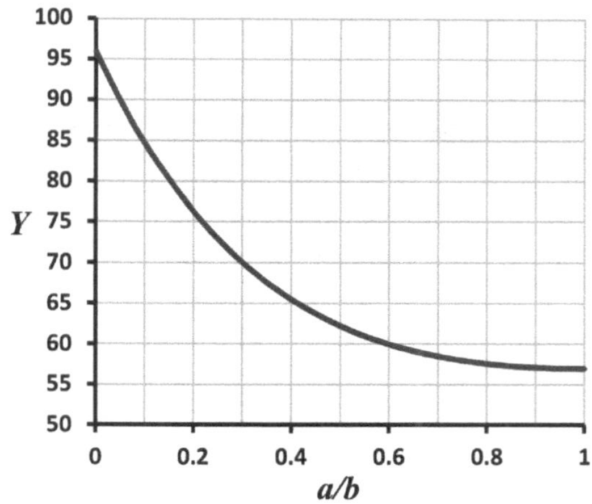

Fig. 12.4 The *Y*-function for annular channels.

12.3.3 The rotational flows

Many more opportunities for solving the equations of motion in terms of stresses are offered in those flow situations in which there is no influence of pressure and of hydrostatic height, that is, in those cases in which the assumption $\nabla P=0$ is met. The equation of motion will then assume the form

$$0=\nabla\cdot\boldsymbol{\tau} \tag{12.45}$$

that can be further simplified for some geometrically simple tasks. In order to preclude a zero motion, it is assumed that one of the planes delimiting the system engages in a rotational motion with a known angular velocity. Most frequently, while one boundary rotates, the other is at rest.

12.3.3.1 Rotation of two coaxial cylinders

The most common case is the flow generated by the rotation of a cylindrical surface. This is a standard case studied most frequently because of its connection to the rotational viscometric flow (see Section 7.3). All what will be left from Eq. (12.17) will be a relation

$$0=\frac{1}{r^2}\frac{\partial\left(r^2\tau_{r\varphi}\right)}{\partial r} \tag{12.46}$$

in which the stress $\tau_{r\varphi}$ is a function of the variable r. The integration of Eq. (12.46) will yield

$$\tau_{\varphi r} = \frac{C}{r^2} \tag{12.47}$$

Alternatively, with the use of the Navier-Stokes equation (10.7), a new derivative will result

$$\frac{\partial}{\partial r}\left(\frac{1}{r}\frac{\partial}{\partial r}\left(ru_{\varphi}\right)\right) = 0 \tag{12.48}$$

The solution of (12.47) has been presented by relations (7.11)–(7.15).

An asymptotic relation for a narrow slot was already dealt with in this text earlier (Eq. 7.17).

12.3.3.2 Rotation of two round plates

This geometry is schematically shown in Fig. 7.4B. The component of the tangential stress $\tau_{z\varphi}$ is a function of the variable z so that what is left from Eq. (12.17) is

$$0 = \frac{\partial \tau_{z\varphi}}{\partial z} \tag{12.49}$$

By performing the integration, one gets

$$\tau_{z\varphi} = C(r) \tag{12.50}$$

and with the use of viscosity the following equation is reached

$$\frac{\partial u_{\varphi}}{\partial z} = \frac{C(r)}{\mu} \tag{12.51}$$

From the boundary conditions

$$\begin{aligned} u_{\varphi} &= \Omega_1\, r \text{ for } z = z_1 \\ u_{\varphi} &= \Omega_2\, r \text{ for } z = z_2 \end{aligned} \tag{12.52}$$

It follows that

$$C(r) = \frac{(\Omega_2 - \Omega_1)\, r}{z_2 - z_1} \tag{12.53}$$

After the integration, it will be

$$\omega \equiv \frac{u_{\varphi}}{r} = \frac{\Omega_2(z - z_1) + \Omega_1(z_2 - z)}{(z_2 - z_1)} \tag{12.54}$$

Usually, one of the disks is immobile so that $\Omega_2 \equiv \Omega$ and $\Omega_1 = 0$. When the origin of the coordinates is placed in the plane $z_1 = 0$, then

$$\omega \equiv \frac{u_\varphi}{r} = \Omega \frac{z}{z_2} \tag{12.55}$$

The torque moment on a disc having the radius R will be

$$M \equiv \int_0^R 2\pi r^2 \mu \frac{du_\varphi}{dz} dr = \frac{\pi}{2} \frac{\mu \Omega R^4}{h} \tag{12.56}$$

where the separation between the two discs is $h \equiv (z_2 - z_1)$.

12.3.3.3 Rotation of two coaxial cones

This geometry is schematically depicted in (Fig. 12.3); its rheometry application has been shown in Fig 7.4D and the special case of a cone on disc in Fig. 7.4C. Both these setups are best treated in spherical coordinates where there is only one nonzero velocity component $v_\varphi(\theta)$ in the direction of rotation and one nonzero component of the stress $\tau_{\theta\varphi}(\theta)$, that is, dependent on only one variable θ. With the use of Eq. (12.21), it will be

$$0 = \frac{1}{s} \frac{d\tau_{\theta\varphi}}{d\theta} + \frac{2 \cos \theta}{s \sin \theta} \tau_{\theta\varphi} \tag{12.57}$$

which yields after integration

$$\tau_{\varphi\theta} = \frac{C_1}{\sin \theta} \tag{12.58}$$

The relevant torque moment between the two cones having a side S can be determined by performing the integration

$$M = \int_0^s \int_0^{2\pi} \tau_{\varphi\theta} s^2 \sin^2\theta \, d\varphi \, ds = \frac{2\pi}{3} C_1 \sin\theta S^3 \tag{12.59}$$

Generally, it holds for all liquids having a constant viscosity, that

$$\tau_{\varphi\theta} = \mu \left[\frac{\sin \theta}{s} \frac{\partial}{\partial\theta} \left(\frac{u_\varphi}{\sin \theta} \right) + \frac{1}{s \sin \theta} \frac{\partial u_\theta}{\partial\varphi} \right] \tag{12.60}$$

where the second term in the brackets is equal to zero. The local angular velocity will be, after the integration,

$$\omega \equiv \frac{u_\varphi}{s\sin\theta} = \frac{\tau_{\varphi\theta}}{\mu}\int\frac{\mathrm{d}\theta}{\sin\theta}\frac{3M}{2\pi\mu S^3\sin^2\theta}\int\frac{\mathrm{d}\theta}{\sin\theta}$$
$$= \frac{3M}{4\pi\mu S^3}\frac{1}{\sin^2\theta}\ln\left(\frac{1-\cos\theta}{1+\cos\theta}\right) + C_2 \tag{12.61}$$

The integration constants can be determined from

$$\omega = \Omega_1 \text{ for } \theta = \theta_1 \tag{12.62}$$
$$\omega = \Omega_2 \text{ for } \theta = \theta_2 \tag{12.63}$$

so that the angular velocity will be given by

$$\omega = \Omega_1 + (\Omega_2 - \Omega_1)\frac{\frac{1}{\sin^2\theta}\ln\left(\frac{1-\cos\theta}{1+\cos\theta}\right) - \frac{1}{\sin^2\theta_1}\ln\left(\frac{1-\cos\theta_1}{1+\cos\theta_1}\right)}{\frac{1}{\sin^2\theta_2}\ln\left(\frac{1-\cos\theta_2}{1-\cos\theta_2}\right) - \frac{1}{\sin^2\theta_1}\ln\left(\frac{1-\cos\theta_1}{1-\cos\theta_1}\right)} \tag{12.64}$$

and the torque

$$M = \frac{4\pi\mu s^3}{3}\frac{\Omega_2 - \Omega_1}{\frac{1}{\sin^2\theta_2}\ln\left(\frac{1-\cos\theta_2}{1+\cos\theta_2}\right) - \frac{1}{\sin^2\theta_1}\ln\left(\frac{1-\cos\theta_1}{1+\cos\theta_1}\right)} \tag{12.65}$$

The angular gap between the cone and the disc in commercial viscometers $\Delta\Theta$ is rather small so that $\theta_1 \equiv \pi/2$ and $\theta_2 \to \pi/2 - \Delta\theta$. Because it is the cone that rotates, it is $\Omega_1 = 0$ and $\Omega_2 \equiv \Omega$ and all the above relations will get simplified to

$$\omega \approx \Omega\frac{\pi/2 - \theta}{\Delta\theta} \tag{12.66}$$

and the momentum transported by the disc of radius $s = R$ will be

$$M \approx \frac{2\pi R^3\mu\Omega}{3\Delta\theta} \tag{12.67}$$

12.3.3.4 Rotation of two coaxial spheres

It would appear at the first glance that this system belongs also in the category of simple viscometric flows, but it turns out that it is not possible to separate or combine components $\tau_{\theta\varphi}$ and $\tau_{s\varphi}$. Yet it is possible to obtain the distribution of the angular velocities by solving the Navier-Stokes equations making further assumptions, such as the assumption of a constant viscosity (Newtonian liquids).

12.4 Two-dimensional unidirectional flows

As was shown in Figs. 12.1 and 12.5, even a small change of the geometric configuration makes the problem two-dimensional. In particular, if the boundary conditions cannot be placed in the characteristic planes of any simple coordinate system that has a high degree of symmetry, then it might be impossible to transpose the problem in a task solvable in stresses. The distribution of the only significant velocity component will depend on the position so that the solution would require the knowledge of the rheological behavior of a given fluid. This can be obtained for the Newtonian fluids by solving the Navier-Stokes equations.

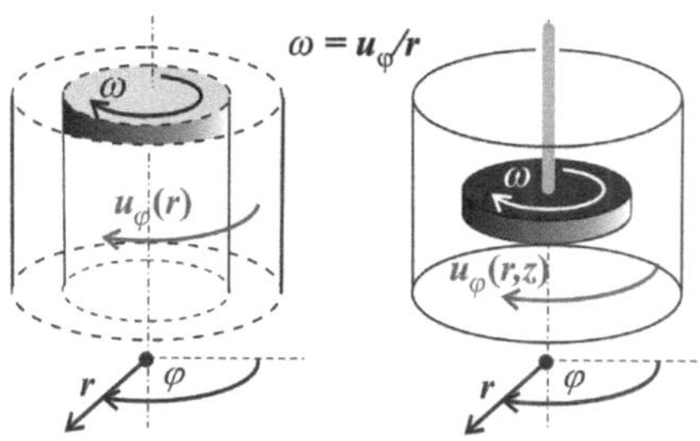

Fig. 12.5 Unidirectional rotational (φ) flow. One-dimensional (r); two-dimensional (r, z).

12.4.1 A channel of rectangular cross section

The geometrical setup depicted in Fig. 12.1 is described by an equation

$$\frac{\partial^2 u_z}{\partial x^2}+\frac{\partial^2 u_z}{\partial y^2}=-\frac{1}{\mu}\frac{\mathrm{d}P}{\mathrm{d}z} \tag{12.68}$$

Because of the symmetry of a rectangular channel, it will be sufficient to solve this equation in just one quarter of the channel for two variables $u_z(x,y)$ which meet the boundary conditions:

$$u_z=0 \text{ for } x=a \text{ at } |y|\le b \tag{12.69}$$

$$u_z=0 \text{ for } y=b \text{ at } |x|\le a \tag{12.70}$$

where a and b are halves of the sides of the rectangle. Another condition to be met is an integral condition that is linked to the average velocity U

$$\int_0^a \int_0^b u_z(x, y) dx dy = Uab \tag{12.71}$$

A complete solution is a field of velocities $u_z(x,y)$ that can be obtained by a numerical solution (see Section 21.1) of the Navier-Stokes equation by one of the commercially available software packages.

Approximate analytical solutions use an expansion of the function $u_z(x,y)$ into a series and subsequent determination of the appropriate coefficients by the optimization of the entropy production by methods of the variation calculus.

The tangential stresses on the surfaces are

$$\tau_{zx}(a) = \mu \left(\frac{\partial u_z}{\partial x}\right)_{x=a} \quad \text{for } x = a \text{ at } |y| \le b \tag{12.72}$$

$$\tau_{zy}(b) = \mu \left(\frac{\partial u_z}{\partial y}\right)_{y=b} \quad \text{for } y = b \text{ at } |x| \le a \tag{12.73}$$

The average tangential stress on the surface is

$$\overline{\tau} \equiv \int_0^a \tau_{zy}(b)\, \mathrm{d}x + \int_0^b \tau_{zx}(a)\, \mathrm{d}y = \left(\frac{-dP}{dz}\right) a\, b \tag{12.74}$$

and the pressure loss is equal to

$$\left(\frac{-dP}{dz}\right) ab = (a + b)\overline{\tau} \tag{12.75}$$

With the use of the equivalent diameter which is for a rectangular cross section $2a \times 2b$,

$$d_e = \frac{4ab}{(a + b)} \tag{12.76}$$

and using the definitions (12.9) and (12.42) the friction coefficient f will be

$$f\, Re \equiv \frac{8\, d_e^2}{\mu\, U}\left(\frac{-dP}{dz}\right) = Y\left(\frac{a}{b}\right) \tag{12.77}$$

Fig. 12.6 shows graphically the asymptotic value of $Y=96$ for a flat channel while for a square cross section of the Y function it is $Y=56.92$. This is a value slightly smaller than 64 for a round tube of the same d_e.

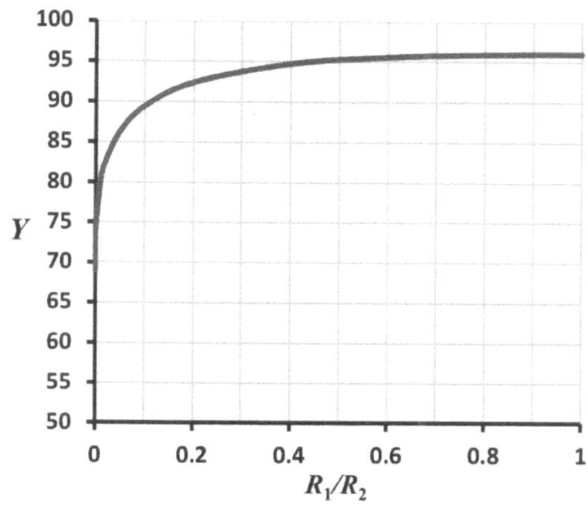

Fig. 12.6 The *Y*-function for rectangular channels.

12.5 The principle of the minimal entropy rise

If the system in which a transport takes place is bounded by a set of independently given boundary conditions, then the process will automatically assume a path of minimum resistance. This is equivalent to assuming a path of minimum entropy growth. In the case of flow this means an ε irreversible dissipation in heat. If, according to Eq. (8.14), the dissipated energy in a unit of volume is then the creation of entropy

$$\frac{dS}{dt} = \int \frac{\epsilon}{T} dV \tag{12.78}$$

In the case of the rectangular channel the local dissipation is

$$\epsilon = \mu \left[\left(\frac{\partial u_z}{\partial x} \right)^2 + \left(\frac{\partial u_z}{\partial y} \right)^2 \right] \tag{12.79}$$

As long as the process goes on at a constant temperature T and viscosity μ, then to meet the requirement of the minimal entropy creation it will suffice if relationship (12.79) holds

$$\int_0^b \int_0^a \left[\left(\frac{\partial u_z}{\partial x} \right)^2 + \left(\frac{\partial u_z}{\partial y} \right)^2 \right] dxdy = \min \tag{12.80}$$

The mathematical discipline that engages in seeking the minimum of a functional (that is, the values of a function in a limited interval) is known as the variation calculus.

Another option is an empirical procedure based on an arbitrary choice of function $u_z(x, y)$, Closer to the correct solution will be the one the functional (12.80) of which is smaller. Alternatively, the best fit could be a linear combination of several such functions. A very good approximation of the correct solution can be obtained relatively easily by such a strategy.

12.6 Instability of the creeping flow

The Reynolds number is not exactly equal to zero even for the creeping flows since some influence of the inertial forces may persist. Even in the case of some curvilinear flows that were earlier classified as unidirectional it is possible to observe, over a certain value of the Reynolds number, deviations of what should happen for a Reynolds number approaching zero. Though the flow is still laminar, the inertial terms (the left side of the Navier-Stokes equation) can no longer be neglected. Two such cases will be here demonstrated:

12.6.1 The instability of flow in a curved channel

If a pipeline of a circular cross section is bent only slightly, that is, if the radius of curvature A is by an order of magnitude larger than the diameter of the tube, then it might be assumed that the velocity distribution will be parabolic as it is in a straight tube.

The basic difference between the flow in a straight tube and a curved tube is in the tendency of the liquid flowing in the center of the tube at the speed U_{max} to move uniformly straight forward (because it has a higher inertia) than the slower liquid at the wall. According to Eq. (10.13) the centrifugal force acting on a unit volume $\varrho U_{max}^2/A$ is given by the radius of curvature A and the angular velocity $\omega = U_{max}/A$. This force causes a slow shift from the center of the tube to the outer side of the curvature by a secondary circulation velocity U_{cir}.

Fig. 12.7 shows the transverse secondary circulation in the cross section of the pipeline. This circulation is slowed down by the viscous forces that are proportional to the expression $\mu\ (U_{cir}/R)$. It follows from the equilibrium of forces in a unit section of the pipe that the characteristic velocity of the circulation U_{cir} is proportional to

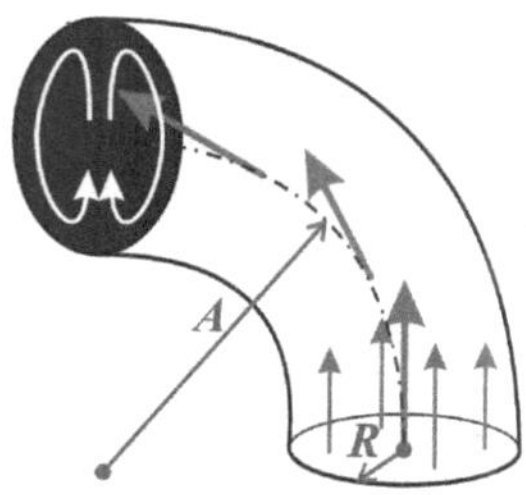

Fig. 12.7 The fast flow in a curved pipe. Centrifugal secondary flow—in white.

$$\frac{U_{cir}}{U} \sim Dn^2 \tag{12.81}$$

where the Dean number, Dn, is defined as

$$Dn \equiv \frac{\rho U d}{\mu} \left(\frac{R}{A} \right)^{1/2} \tag{12.82}$$

Although the secondary circulation occurs already during the creeping flow, it becomes significant, according to the above proportion, only at $Dn > 10$. Much more energy is lost due to the secondary circulation than in a straight tube. This becomes obvious from the relationship

$$f \equiv \frac{64}{Re} \left(1 + 0.028\, Dn^{0.65} \right) \tag{12.83}$$

12.6.2 The instability of rotational flows

A body having a characteristic radius R that slowly rotates at an angular velocity Ω induces the surrounding fluid into a rotation that has only one component of flow in the direction of φ. In the space where the angular velocity in the direction away from the axis drops (outside of the rotor), the centrifugal force causes a centrifugal flow (schematically shown by the dashed lines in Fig. 12.8).

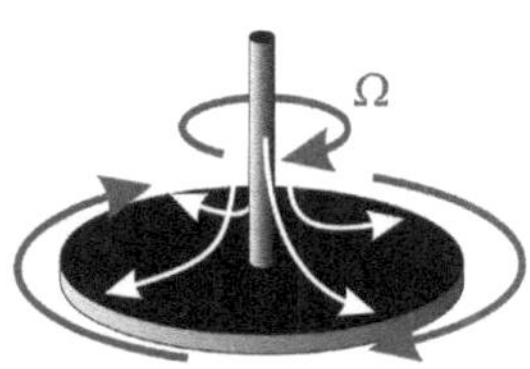

Fig. 12.8 The fast flow around a rotating body. Centrifugal secondary flow—in white.

Analogous considerations about the balancing of the centrifugal force by the viscous resistance lead to an estimate that the characteristic velocity of the secondary centrifugal flow U_r will be proportional to the second power of the Reynolds number

$$\frac{U_r}{U} \sim Re^2 \tag{12.84}$$

where the Reynolds number for rotation is defined as

$$Re \equiv \frac{\rho \Omega R^2}{\mu} \tag{12.85}$$

In the engineering practice (particularly when dealing with mixing) it is common to use for the Reynolds number the diameter of the rotor $d = 2R$ as the characteristic longitudinal dimension. Also, the frequency of rotation $N = \Omega/(2\pi)$ is used instead of the angular velocity. The Reynolds number for rotation is then equal to

$$Re_N \equiv \frac{\rho N d^2}{\mu} = \frac{2}{\pi} Re \tag{12.86}$$

The secondary flow contributes to the growth of the torque moment mainly at high Reynolds numbers. These relationships differ for different rotors.

The secondary flow is undesirable when measuring viscosity by rotational viscometers but it is welcome in mixing processes.

A special case of the secondary centrifugal motion occurs when a cylindrical spindle rotates in a cylindrical case (Fig. 12.5). In this setup the centrifugal flow forms circular loops of finite dimension—roughly equal to the width of the gap between the stator and the rotor—known as the Taylor vortices. They are depicted in Fig. 12.9.

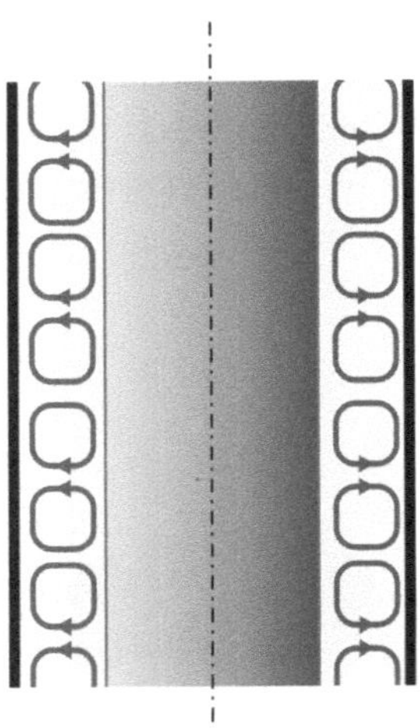

Fig. 12.9 The Taylor vortices between a rotating cylinder and a cylindric vessel.

12.7 Summary

There are several situations for which it is possible to state that, in some coordinate systems, the steady creeping flow runs only in one direction: unidirectional flows. Among these, some are one-dimensional, which means that the velocity depends only on one coordinate. It was shown in Chapter 7 that these flows can be analyzed by individually derived relationships. However, it is possible to reach a mathematical description also by simplifying the Cauchy equations. In this way, it is possible to arrive at relatively simple solvable ordinary differential equations for the distribution of stresses and, for a known viscous behavior, also for the distribution of velocities.

For more complex flows (such as a unidirectional flow depending on two coordinates) it is necessary to use the Navier-Stokes equations that, even after the elimination of the identically zero terms, lead to partial differential equations. In most cases, these are equations that can be solved numerically without much problem by commercially available software.

Sometimes it is possible to make a rough estimate of how the velocity profile should look like and suggest an approximation either by an empirical function or by a series. The principle of a minimal entropy growth, formalized by the methods of the variation calculus, can be used to select the most likely approximation. Alternatively, the numerical Monte Carlo methods (Section 21.1) can be used.

When physically realizing the creeping flows, it is necessary to keep in mind the influence of the inertial forces. While the inertial forces stabilize the flows in the straight channels (be it of a circular or noncircular cross section) so that disturbances (turbulence) do not occur until the Reynolds number reaches over 1000, in curvilinear flows the inertial forces may bring about regular secondary flow at Reynolds numbers in the vicinity of one.

CHAPTER THIRTEEN

The basics of steady heat conduction and diffusion

Contents

13.1 The three-dimensional Fourier's law

The fundamental relation defining the **conduction of heat** in an immobile material having the **thermal conductivity** k was formulated earlier in this text in a coordinate system chosen specifically for this given case. This equation, however, needs to be generalized to be valid in any coordinate system. Fig. 13.1 and Eq. (13.1) will depict the relationship between the **intensity of heat flow** from a plane having a **temperature** T_1 to another plane removed from the first plane by a distance h, rotated by an angle α around the z-axis, and maintained at a temperature T_2

$$q = -k\frac{T_2 - T_1}{h} \tag{13.1}$$

The heat flow in the case of a unidirectional heat conduction depicted in the Cartesian coordinate system if the plane is partially inclined (Fig. 13.1).

Transport and Surface Phenomena
https://doi.org/10.1016/B978-0-12-818994-8.00013-0

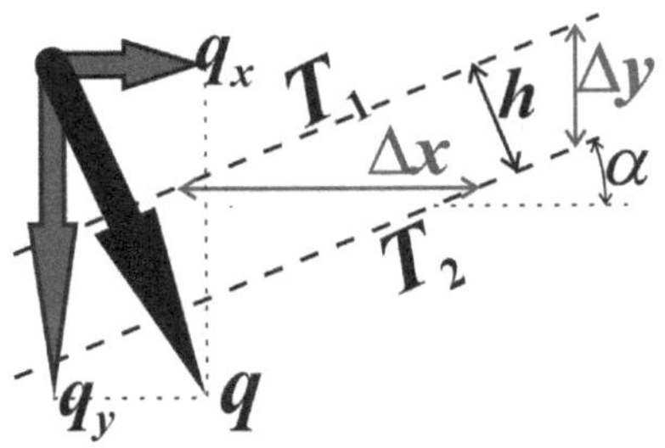

Fig. 13.1 The heat flow in the case of a unidirectional heat conduction depicted in the Cartesian coordinate system if the plane is tilted.

The vector of the intensity of heat flow $\boldsymbol{q}$ can be divided into two components

$$q_x = q \sin \alpha \, q_y = q \cos \alpha \tag{13.2}$$

and similarly it will be

$$\frac{\partial T}{\partial x} = \frac{T_2 - T_1}{h/\sin \alpha} \, \frac{\partial T}{\partial y} = \frac{T_2 - T_1}{h/\cos \alpha} \tag{13.3}$$

so that

$$q_x = -\lambda \frac{\partial T}{\partial x} \, q_y = -\lambda \frac{\partial T}{\partial y} \tag{13.4}$$

It can be shown that an analogical result would be obtained if the set was inclined against the z-axis. This result can be generalized for any coordinate system if using the vector of the intensity of heat flow $\boldsymbol{q}$ and the temperature gradient $\boldsymbol{\nabla} T$

$$\boldsymbol{q} = -\lambda \boldsymbol{\nabla} T \tag{13.5}$$

thus arriving at the **generalized Fourier law of heat conduction**.

13.2 Three-dimensional Fick's law

A similar case is the **diffusion** of the minor component A in an immobile medium in a material that has for the other component a **diffusivity** $\mathcal{D}$ [m^2/s]. The **intensity of the mass rate** $\boldsymbol{j_A}$ is defined as the amount of the mass of the component passing through a unit area during a unit of time [mol/(m^{2}s)]. It can be shown that the relationship between the vector of the intensity of mass flow $\boldsymbol{j_A}$ and the field **of molar concentration** C_A will be

$$j_A = -\mathcal{D}\nabla C_A \tag{13.6}$$

where ∇C_A is the **concentration gradient**. This equation is known as the general **Fick's law**.

In some cases, it is more advantageous to balance the mass of the component A; **the intensity of mass flow** $\boldsymbol{n_A}$ is the mass of the component passing through a unit area during the unit of time [the basic unit being $kg/(m^2s)$]. If the **mass concentration** gets the symbol c_A [the basic unit being kg/m^2], then the equation for the diffusion will acquire the same formula

$$n_A = -\mathcal{D}\nabla c_A \tag{13.7}$$

known as the **generalized Fick's law of diffusion**. It needs to be reemphasized that the diffusivity $\mathcal{D}$ is the same in both cases (in terms of the dimension as well as of the value).

13.3 The heat conduction

The thermodynamics asserts that there is a law of energy preservation. Energy can have miscellaneous manifestations: one of them is heat. When performing an energy balance, it is necessary to keep in mind that miscellaneous forms of energy can change in heat (source) and that the heat energy can change in another forms of energy (heat loss). Very often these changes are insignificant. Therefore, it is possible to assume, with an acceptable accuracy, the preservation of heat energy. Eventually, a term covering the influence of a heat source or a heat loss can be added. Because the pressure in the systems under consideration changes only insignificantly the heat energy is essentially identical with enthalpy. In accordance with thermodynamics, the heat energy is related to some reference temperature, usually to the zero temperature in the used temperature scale.

The instantaneous amount of heat energy (enthalpy) in a unit volume of a continuum is

$$Q = \int_0^T \rho\, c_P\, dT \tag{13.8}$$

In a volume element $\Delta V = \Delta x\ \Delta y\ \Delta z$ of a continuum located at the x, y, z coordinate of a Cartesian coordinate system, the **accumulation of the heat energy** is given by the difference between the final value of the heat energy

$$Q(x, y, z, t+\Delta t)\ \Delta x \Delta y \Delta z$$

and its initial value

$$Q(x, y, z, t)\ \Delta x \Delta y \Delta z$$

13.3.1 Detailed derivation

The boundary layers contribute to the conduction of heat by an amount $\boldsymbol{q}$ that is delivered over the time Δt from the region of lower x

$$q_x(x, y, z, t)\ \Delta y \Delta z \Delta t$$

while the outflow into the region of higher x is

$$q_x(x+\Delta x, y, z, t)\ \Delta y \Delta z \Delta t$$

A similar process takes place in the y- and z-directions.

In summary, the balance of the heat conduction will be

$$\begin{array}{c} +Q(x, y, z, t+\Delta t)\ \Delta x \Delta y \Delta z \\ -Q(x, y, z, t)\ \Delta x \Delta y \Delta z \\ +q_x(x\Delta x, y, z, t)\Delta y \Delta z \Delta t \\ +q_y(x, y+\Delta y, z, t)\Delta y \Delta z \Delta t \\ +q_z(x, y, z+\Delta z, t)\ \Delta y \Delta z \Delta t \\ -q_x(x, y, z, t)\ \Delta y \Delta z \Delta t \\ -q_y(x, y, z, t)\ \Delta x \Delta z \Delta t \\ \underline{-q_z(x, y, z, t)\ \Delta x \Delta y \Delta t} \end{array} = 0 \tag{13.9}$$

13.3.2 The result

By dividing all terms by a product $\Delta x\ \Delta y\ \Delta z\ \Delta t$ and taking account of the definition of derivation, a differential relation is obtained

$$\frac{\partial Q}{\partial t} + \frac{\partial q_x}{\partial x} + \frac{\partial q_y}{\partial y} + \frac{\partial q_z}{\partial z} = 0 \tag{13.10}$$

Making use of the definition of the heat energy (Eq. 13.8) and of the Fourier law (Eq. 1.3), the **equation of heat conduction** in terms of temperatures is obtained as

$$\frac{\partial(\rho c_P T)}{\partial t} = -k\left(\frac{\partial^2 \mathrm{T}}{\partial \mathrm{x}^2} + \frac{\partial^2 T}{\partial y^2} + \frac{\partial^2 T}{\partial z^2}\right) \tag{13.11}$$

With the use of the Laplace operator, Eq. (12.20) can be reduced to

$$\frac{\partial(\rho c_P T)}{\partial t} = -k\,\Delta^2 T \tag{13.12}$$

In the most common case, when the material properties ρ and c_P do not change significantly, the simplified equation of heat conduction

$$\rho\, c_P \frac{\partial T}{\partial t} = -k\,\Delta^2 T \tag{13.13}$$

will suffice. Sometimes the thermal diffusivity $a \equiv k/(\rho \cdot c_P)$ (Eq. 2.14) is preferred to write

$$\frac{\partial T}{\partial t} = a\,\nabla^2 T \tag{13.14}$$

13.3.3 The heat source

One of the tasks that occur is the case involving a heat source. A simple example is the uniform positive heat source that adds to each unit volume over a unit time period a positive (or negative) heat energy $\boldsymbol{\Phi}$ ($\mathrm{W/m^3}$). The source could be electric heating, exothermal (or endothermic) chemical reaction, absorption (or radiation) of some kinds of rays, reversible change of mechanical energy by compression (and a negative heat source due to an expansion), or an irreversible dissipation of mechanical energy in heat due to friction or viscosity.

In more complex cases, the value of $\boldsymbol{\Phi}$ may depend on temperature. However, in this case, it will be assumed that it is known. It adds to the accumulation of heat and thus, it complements substance flow the Fourier-Kirchhoff equation to become

$$\rho\, c_P\left(\frac{\partial T}{\partial t} + \boldsymbol{u}\cdot\boldsymbol{\nabla} T\right) = -k\Delta^2 T + \boldsymbol{\Phi} \tag{13.15}$$

13.4 The equation of the mass transport

Because the mass composition, just like the temperature, is a scalar quantity the transport of mass and heat can be described by formally similar relationships it will not be necessary to deal with them in a great detail.

13.4.1 Diffusion

The accumulation of substance A in a unit volume can be expressed as a change of concentration C_A over a unit of time, thus

$$\frac{\partial C_A}{\partial t}$$

In the case of a pure diffusion, the change of the concentration is due to the sum of the substance flow vectors $\boldsymbol{j}''$ into the element and out.

Using the same procedure as in the case of heat conduction, the equation of diffusion will be

$$\frac{\partial C_A}{\partial t} = -\mathcal{D}\left(\frac{\partial^2 C_A}{\partial x^2} + \frac{\partial^2 C_A}{\partial y^2} + \frac{\partial^2 C_A}{\partial z^2}\right) \tag{13.16}$$

A shorter expression will be obtained by the use of the Laplace operator

$$\frac{\partial C_A}{\partial t} = -\mathcal{D}\Delta^2 C_A \tag{13.17}$$

13.4.2 The convective mass transport

If the diffusion takes place in a moving medium, then the equation must be supplemented by a term expressing the transport of mass by the flow of the medium and the equation of diffusion becomes

$$\begin{aligned}\frac{\partial C_A}{\partial t} + u_x\frac{\partial C_A}{\partial x} + u_y\frac{\partial C_A}{\partial y} + u_z\frac{\partial C_A}{\partial z}\\ = -\mathcal{D}\left(\frac{\partial^2 C_A}{\partial x^2} + \frac{\partial^2 C_A}{\partial y^2} + \frac{\partial^2 C_A}{\partial z^2}\right)\end{aligned} \tag{13.18}$$

With the use of the vector calculus, the **equation of the convective mass transport** becomes

$$\frac{\partial C_A}{\partial t} + \boldsymbol{u}\cdot\nabla C_A = -\mathcal{D}\Delta^2 C_A \tag{13.19}$$

13.4.3 A substance change

Although the substance changes by a chemical reaction is a frequent process, it is beyond the scope of this chapter. It is dealt within the engineering of chemical reactors which is a separate scientific discipline.

13.5 Qualitative aspects of transport phenomena

When pondering about transport phenomena then the first, and often the most important part of the reflection is to define the problem qualitatively. Only then it becomes sensible to choose a mathematical expression of the problem and solve it. Therefore, it is useful to study highly simplified tasks the solution of which will indicate the role of transport phenomena and the likely outcome of the process.

In the following, it will be shown how the temperature field will look like when the surface of a cold body (in *blue*) is brought to a contact with a warmer surface (in *red*).

First, the gradual cooling of the medium will take place due to the **heat conduction** in the close vicinity of the cold surface (Fig. 13.2).

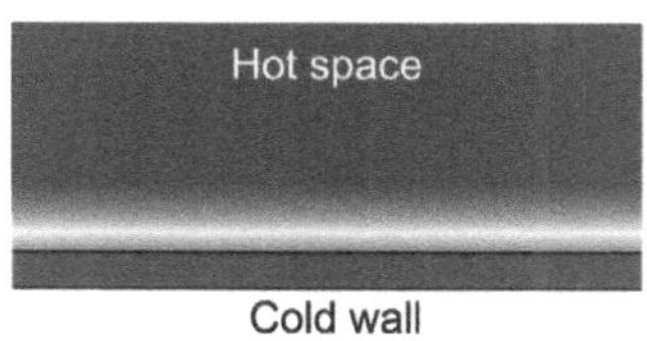

Fig. 13.2 A developing temperature field.

The layer of the half-space affected by the conduction will later become wider (Fig. 13.3).

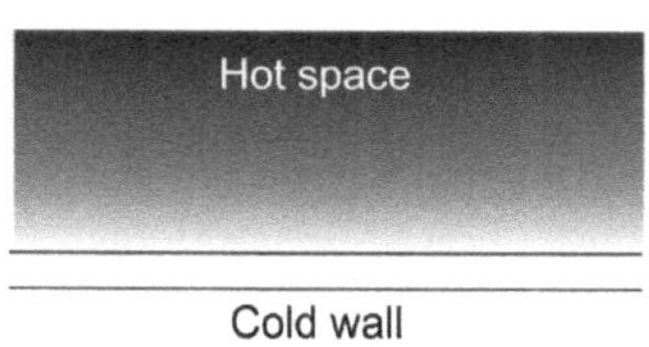

Fig. 13.3 A developing temperature field later.

Sometimes the warm space might be bordered by a warm wall; then after a while, the temperature distribution will become stable (Fig. 13.4).

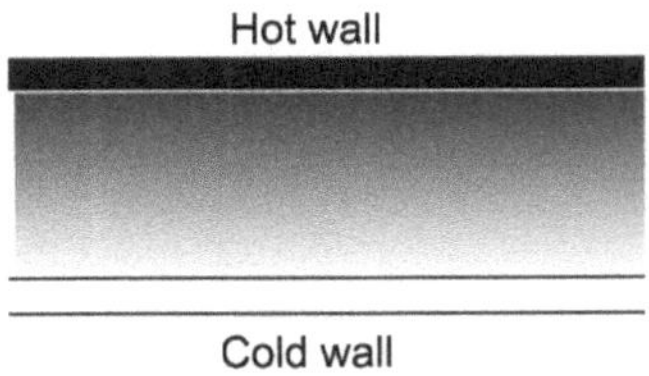

Fig. 13.4 The developed temperature field.

Should the cold surface have a finite dimension, then **boundary effects** might appear that might, but need not become significant (Fig. 13.5).

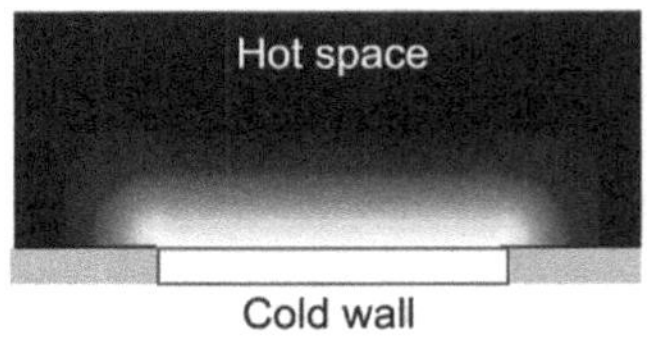

Fig. 13.5 The side effects above the finite chilled surface.

The boundary effects will assume a key importance if the chilled space is substantially larger than the cold surface and the time is long enough. The temperature distribution then acquires a significantly different geometry that, depending on circumstances, might show a point or a circular symmetry (Fig. 13.6).

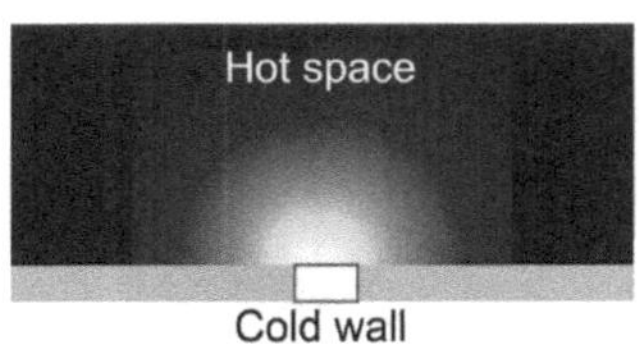

Fig. 13.6 The temperature field above a small chilled surface.

The temperature distribution will keep changing if the space around the cooler wall moves. Then, the region of altered temperatures will widen in the direction of the flow. The contribution of the **heat convection** will increase. This will be a **forced convection** because the motion is triggered by an outside interference (Fig. 13.7).

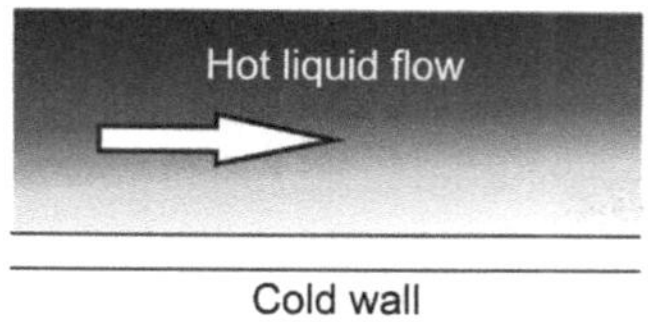

Fig. 13.7 The laminar forced convection.

A more vigorous flow of fluids is often accompanied by a randomly spread whirling (turbulence) that causes a transverse as well as a longitudinal mixing thus accelerating the heat convection (Fig. 13.8).

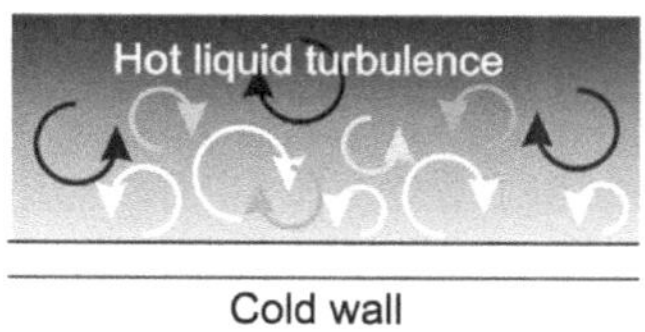

Fig. 13.8 The turbulent heat transfer.

In some cases, the irregular distribution of temperatures in the liquid medium leads to the variation of the density. Normally, the colder layer has a higher density (among the few exceptions is water at temperatures 0–4°C). Usually, the fluid having higher density has a tendency to take a position with a lower potential. (Customarily, but not quite correctly it is said that the warmer air or water rise in a gravitation field.) The flow triggered by such forces is called the natural or the free convection (Fig. 13.9).

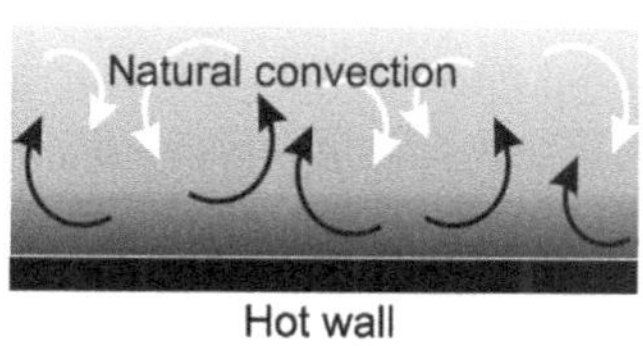

Fig. 13.9 The natural (or the free) convection.

Positioning of the colder surface up will bring about instability in the fluid manifested by the generation of more or less regular circular loops that spread the heavier fluid down and the lighter fluid up. The case of a rotating fluid where the heavier fluid is pushed away from the axis of rotation due to the centrifugal force is somewhat different. In some cases, the natural convection may assume a chaotic turbulent motion leading to the formation of roughly regular dissipative structures of rising and falling zones (Fig. 13.10).

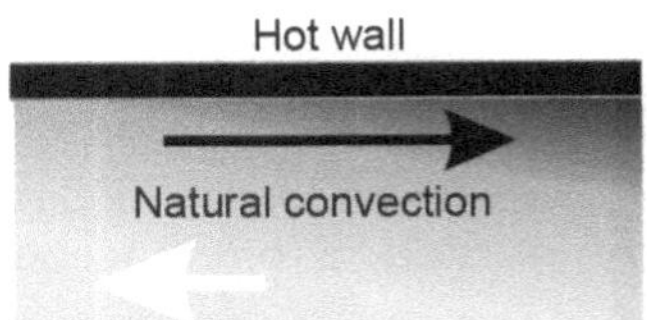

Fig. 13.10 The dissipative structures for the free convection.

The study of transport phenomena makes it possible to formulate for specific cases suitable mathematical models and seek appropriate solutions for them.

Although the progress in numerical mathematics and in the computational methods makes it possible to solve increasingly more complex problems, it is useful to focus on solving **simpler problems** that provide a **graphical picture** of the processes going on in the system. Also, they allow to assess the possibility of neglecting the less important variables without jeopardizing the accuracy of the result.

When attempting to involve all variables simultaneously, then the subject becomes confused. The chance of finding an analytical solution is lost. Moreover, there is a danger that the numerical solution might generate an unspecified, possibly a fatal error. In the best case the result is a **case study,** meaning to get just **one field for one instant.** An infinite number of such solutions would be necessary to understand the whole action. Further, it would be necessary to analyze **the parametric sensitivity** of the result to the individual given variables.

13.6 Summary

The alignment of scalar variables, such as the temperature or concentration, can be expressed by vector relationships for the transport of heat and mass.

The transfer of heat in an immobile medium can be expressed by the generalized Fourier law of heat conduction that formulates a linear relationship between the vector of heat flow and the vector of temperature gradient. The coefficient of proportionality is the negative value of the thermal conductivity. The heat energy balance can be expressed as the equation describing the temperature distribution in space.

The transfer of mass in space is formally similar to the transport of heat. The generalized Fick's law of diffusion introduces a linear proportionality between the vector of the mass flow and the vector of the concentration gradient of the substance. The coefficient of proportionality, in both cases, is the diffusivity.

The balance of the heat energy in a moving medium must have an additional term expressing the influence of heat brought in and out with the moving medium. The resulting expression is the Fourier-Kirchhoff equation which includes the terms for heat accumulation, heat convection, and heat conduction. Additional terms for the conversion of other forms of energy in heat or for a negative source can be added.

When solving the real cases of heat and mass transfer, it is worthwhile to make an estimate whether a substantial portion of the process does not show some symmetry that would allow neglecting any of the directions of the transfer. Any such omission would significantly simplify the mathematical solution and thus provide at least the first, and possibly a satisfactory estimate of the process.

CHAPTER FOURTEEN

Unidirectional unsteady transfer

Contents

14.1 Unidirectional unsteady transfer in half-space

What is left from the Navier-Stokes equations is in Cartesian coordinates just

$$\frac{\partial u_y}{\partial t} = \nu \frac{\partial^2 u_y}{\partial x^2} \tag{14.1}$$

What is left from the Fourier equations is in Cartesian coordinates just an equation

$$\frac{\partial T}{\partial t} = a \frac{\partial^2 T}{\partial x^2} \tag{14.2}$$

Transport and Surface Phenomena
https://doi.org/10.1016/B978-0-12-818994-8.00014-2

Balance species A in Cartesian coordinates is expressed by the equation, called the Fick second law

$$\frac{\partial C}{\partial t} = D\frac{\partial^2 C}{\partial x^2} \tag{14.3}$$

It is obvious that, from the point of view of the mathematical treatment, the transport processes are identical. Formally, they can be described by the same equation

$$\frac{\partial s}{\partial t} = B\frac{\partial^2 s}{\partial x^2} \tag{14.4}$$

where the measure of the particular transport is the scalar s such as the velocity, temperature, or concentration, while the intensity of the transport, such as the stress, the heat, or mass flow are proportional to the kinematic transport coefficient B that could be the kinematic viscosity, thermal, or mass diffusivity. The overview of the particular quantities is given in Table 14.1.

Table 14.1 Analogy of the unsteady unidirectional transfer.

Quantity transferred	Momentum	Heat	Species
Dynamic transport coefficient, b	Dynamic viscosity, μ	Thermal conductivity, k	Diffusivity, $\mathcal{D}$
Kinematic transport coefficient, B	Kinematic viscosity, $\nu \equiv \mu/\rho$	Temperature diffusivity, $a \equiv k/(\rho c_p)$	Diffusivity, $\mathcal{D}$
Scale, s	Velocity, u_y	Temperature, T	Concentration, C_A
Intensity of transfer $-b\frac{\partial s}{\partial x}$	Momentum flow, $-\tau_{yx} = -\mu\frac{\partial u_y}{\partial x}$	Heat flow, $q_x = -k\frac{\partial T}{\partial x}$	Mass flow, $n_x = -\mathcal{D}\frac{\partial C_A}{\partial x}$
Transfer at the wall $-b\frac{\partial s}{\partial x}\Big\vert_{x=0}$	$-\tau_w = -\mu\frac{\partial u_y}{\partial x}\Big\vert_{x=0}$	$q_w = -k\frac{\partial T}{\partial x}\Big\vert_{x=0}$	$n_w = -\mathcal{D}\frac{\partial C_A}{\partial x}\Big\vert_{x=0}$
Gradient $\frac{\partial s}{\partial x}$	Velocity gradient, $\frac{\partial u_y}{\partial x}$	Temperature gradient, $\frac{\partial T}{\partial x}$	Concentration gradient, $\frac{\partial C_A}{\partial x}$

Table 14.1 Analogy of the unsteady unidirectional transfer.—cont'd

Quantity transferred	Momentum	Heat	Species
Gradient at the wall $s'_w = \left.\frac{\partial s}{\partial x}\right\rvert_{x=0}$	Velocity gradient, $u'_w = \left.\frac{\partial u_y}{\partial x}\right\rvert_{x=0}$	Temperature gradient, $T'_w = \left.\frac{\partial T}{\partial x}\right\rvert_{x=0}$	Concentration gradient, $C'_w = \left.\frac{\partial C_A}{\partial x}\right\rvert_{x=0}$
Boundary layer thickness $\frac{s_w(x,t) - s_\infty}{-s'_w}$	Momentum boundary layer $\delta_I \equiv \frac{u_y(0,t) - u_\infty}{-u'_w}$	Thermal boundary layer $\Delta_I \equiv \frac{T(0,t) - T_\infty}{-T'_w}$	Concentration boundary layer $\Delta_I \equiv \frac{C_A(0,t) - C_{A\infty}}{-C'_w}$

14.1.1 Initial and boundary conditions

The solution of a specific problem requires the knowledge of the initial condition and of the boundary conditions. The general initial condition could be any known arbitrary overall distribution

$$s = s_0(x) \quad \text{at } t = 0 \tag{14.5}$$

be it calculated or measured. In the half-space is also considered

$$\frac{\partial s}{\partial x} = 0 \quad \text{and} \quad s \to s_\infty \equiv s_0(\infty) \quad \text{for } x \to \infty \quad \text{at any time } t \geq 0 \tag{14.6}$$

The instant $t = 0$ is usually chosen to coincide with some event, manifested by a **step change of boundary conditions.**

The first kind of the boundary conditions is represented by a **prescribed value *s* at the boundary** (the Dirichlet boundary conditions), i.e.,

$$s = s_w(t) \quad \text{for } x = 0 \quad \text{at any time } t > 0 \tag{14.7}$$

The second kind assumes a prescribed transfer intensity, that is, a **prescribed gradient** $(\partial s/\partial x)_{x=0}$ **at the boundary** (the Neumann boundary conditions), i.e.,

$$\left.\frac{\partial s}{\partial x}\right\rvert_{x=0} = s'_w(t) \quad \text{for } x = 0 \quad \text{at any time } t > 0 \tag{14.8}$$

There is a ***third type*** *of boundary conditions appropriate, for example, for a situation when the interface between the observed system and its surrounding represents a* ***given resistance*** *to the heat conduction similar to the resistance to heat conduction through a solid wall. If it is possible to describe or at least estimate the transfer of heat outside the* *boundary of the observed system, then it should be possible to compare the flow of the heat to the boundary with the calculated heat transfer inside the system and thus evaluate the acceptability of the assumption made about the conditions at the boundary. Mathematically, this can often be formulated as a functional dependence between the temperatures at the boundary with the local gradient. If, however, the resistance at the boundary is very low, then this condition will change in the condition of a known temperature. In the case of a very high resistance of the wall, the resistance allows to penetrate only a well-defined heat flow. The result of such process can be expected somewhere between the asymptotes determined by simpler techniques, for the case of the first or the second boundary condition.*

An event that can mark the beginning of the observed (velocity, temperature, and concentration) development and movement of the heat could also be an instantaneous **change of the equation of transfer**, for example, the rise and the extinction of a magnetic force, volume source, such as a chemical reaction, electrical resistance, induction or microwaves, expansion or compression of a gas, radiation, etc.

In the following paragraphs, simple cases of first or second boundary conditions will be dealt with that are placed in coordinate surfaces of simple Cartesian system. Originating from the assumption of a constant step either of the value s (velocity, temperature, and concentration) or of the gradient (i.e., the step change of either a shear stress, heat flow, or mass flow), respectively.

Some situations will be presented only for the case of the unsteady heat transfer since this is the most frequently occurring case. Clearly, the results can be applied also to the other transports using Table 4.1.

14.2 Unidirectional unsteady transfer in half-space

14.2.1 Typical conditions

A typical **initial condition** is the state reached by the observed process as it was attained in the preceding phase. The simplest and most frequently used condition assigned to the beginning of the observed process at the instant from which the time is counted ($t=0$) is the constant value s in the whole system.

$$s = s_0 \quad \text{at } t = 0 \tag{14.9}$$

The instant $t=0$ is usually chosen to coincide with some event, manifested by a **step change of boundary conditions**.

14.2.2 The first fundamental kind

The boundary conditions are represented by a **step change of the value** s at the wall

$$s = s_w \quad \text{for } x = 0 \quad \text{at any time } t > 0 \tag{14.10}$$

With respect to the boundary, the partial differential equation can be transferred to the ordinary equation by the method of variables combination using the dimensionless coordinate

$$\xi \equiv \frac{x}{2\sqrt{Bt}} \tag{14.11}$$

The transfer equation (14.4) assumes the form

$$-2\xi \frac{\mathrm{d}s}{\mathrm{d}\xi} = \frac{\mathrm{d}^2 s}{\mathrm{d}\xi^2} \tag{14.12}$$

This equation is valid relatively generally. It is, however, of special interest because the above-listed three initial and boundary conditions are reduced to only two

$$s(0) = s_w \tag{14.13}$$

$$s(\infty) = s_0 \tag{14.14}$$

The solution of this problem can be, therefore, expressed as a function of $s(\xi)$. The integration will lead to

$$\frac{\mathrm{d}s}{\mathrm{d}\xi} = -(s_w - s_0) \exp\left(-\xi^2\right) \tag{14.15}$$

and

$$s(x, t) = s_0 + (s_w - s_0) \operatorname{erfc}(\xi) \tag{14.16}$$

where the special function $\operatorname{erfc}(\xi)$ defined by

$$\operatorname{erfc}(\xi) \equiv \frac{2}{\sqrt{\pi}} \int_{\xi}^{\infty} e^{-x^2} dx \tag{14.17}$$

can be found in specialized tables or can be drawn from the Microsoft calculator Excel.

This function starts at erfc(0)c = 1 and converges with growing ξ very fast to zero. For example, erfc(2) ≈ 0.0046, erfc(3) ≈ 0.000022, erfc(4) ≈ 0.000000015.

This is not at variance with the molecular theory of transport phenomena according to which the collisions of the particles occur with a finite speed.

Relation (14.15) results in

$$s'_w \equiv \left.\frac{\partial s}{\partial x}\right|_{x=0} = \frac{-(s_w - s_0)}{\sqrt{\pi B t}} \tag{14.18}$$

The tangent aims at a point that is the characteristic thickness of the boundary layer

$$\delta_I \equiv \frac{s_w - s_0}{-s'_w} \tag{14.19}$$

so that

$$\delta_I = \delta_1(t) = \sqrt{\pi B t} \approx 1.772\sqrt{Bt} \tag{14.20}$$

Alternatively, this could be expressed as a function of the argument (x/δ_1),

$$\frac{s - s_0}{s_w - s_0} \equiv \psi_1\left(\frac{x}{\delta_1}\right) = \operatorname{erfc}\left(\frac{\sqrt{\pi}}{2}\frac{x}{\delta_1}\right) \tag{14.21}$$

depicted in Fig. 14.1. The fields of the individual quantities are expressed by

$$T(x,t) = T_0 + (T_w - T_0)\operatorname{erfc}\left(\frac{x}{2\sqrt{at}}\right) \tag{14.22}$$

$$C_A(x,t) = C_0 + (C_w - C_0)\operatorname{erfc}\left(\frac{x}{2\sqrt{\mathcal{D}t}}\right) \tag{14.23}$$

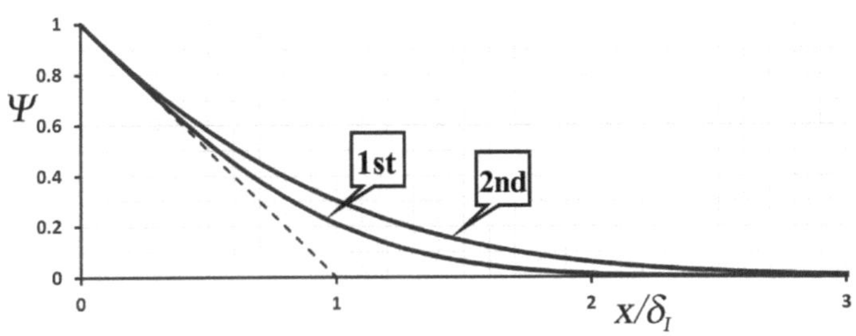

Fig. 14.1 Dimensionless profile in a half-space.

The specific forms of the partial results are summarized in Table 14.2.

Table 14.2 Unsteady unidirectional transfer in half-space; first fundamental kind of the boundary conditions.

Quantity transferred	Momentum	Heat	Species
Initial condition for			
$t=0 \quad s=s_0$	$u_y=u_0$	$T=T_0$	$C_A=C_0$
Boundary condition for			
$x=0 \quad s=s_w$	$u_y=u_w$	$T=T_w$	$C_A=C_w$
Dimensionless coordinate			
$\xi\equiv\frac{x}{2\sqrt{Bt}}$	$\xi\equiv\frac{x}{2\sqrt{\nu t}}$	$\xi\equiv\frac{x}{2\sqrt{at}}$	$\xi\equiv\frac{x}{2\sqrt{\mathcal{D}t}}$
Definition			
$\psi_1\equiv\frac{s-s_0}{s_w-s_0}$	$w_1\equiv\frac{u_y-u_0}{u_w-u_0}$	$\vartheta_1\equiv\frac{T-T_0}{T_w-T_0}$	$\chi_1\equiv\frac{C_A-C_0}{C_w-C_0}$
Dimensionless profile			
$\psi_1=\text{erfc}(\xi)$	$w_1=\text{erfc}(\xi)$	$\vartheta_1=\text{erfc}(\xi)$	$\chi_1=\text{erfc}(\xi)$
Value at the wall	Velocity	Temperature	Concentration
$\psi_1(0,t)=1$	$w_1(0,t)=1$	$\vartheta_1(0,t)=1$	$\chi_1(0,t)=1$
Gradient at the wall	Velocity gradient,	Temperature gradient,	Concentration gradient,
$s'_w=\frac{-(s_w-s_0)}{\sqrt{\pi Bt}}$	$u'_w=\frac{-(u_w-u_0)}{\sqrt{\pi\nu t}}$	$T'_w=\frac{-(T_w-T_0)}{\sqrt{\pi at}}$	$C'_w=\frac{-(C_w-C_0)}{\sqrt{\pi\mathcal{D}t}}$
Boundary layer thickness	Momentum boundary layer	Thermal boundary layer	Concentration boundary layer
$\delta_I\equiv\frac{s_w-s_0}{-s'_w}$	$\delta_{1I}=\sqrt{\pi\nu t}$	$\Delta_{1I}=\sqrt{\pi at}$	$\Delta_{1I}=\sqrt{\pi\mathcal{D}t}$
Transfer at the wall			
$-B\frac{\partial s}{\partial x}\Big\vert_{x=0}$	$-\tau_w=\mu\frac{u_w-u_0}{\sqrt{\pi\nu t}}$	$q_w=k\frac{T_w-T_0}{\sqrt{\pi at}}$	$n_w=\mathcal{D}\frac{C_w-C_0}{\sqrt{\pi\mathcal{D}t}}$

In a specific case, the conditions of the first fundamental kind may be represented by

- **For momentum transfer**: Step change of the wall tangential velocity.
- **For heat transfer**: Contacting the system with some larger block of material of known temperature that would transfer heat well toward the boundary. Such a block could be a well-mixed fluid (e.g., a thermostat) or a solid material that has good thermal conductivity. A good way of maintaining a constant temperature at a boundary is the phase change taking place at a defined boiling temperature (steam heating, condensation) or the melting temperature of a pure material.
- **For mass transfer**: Contacting the system with some larger block of material of known concentration. A constant concentration at a boundary

may be maintained by the phase change like evaporation, condensation, and dissolution of a pure substance. Such a condition can be met also with certain heterogeneous reactions, particularly with an electrolysis under the conditions of limiting diffusion current.

Example:

After a jump change of temperature on the surface of a rock, the thermal diffusivity of which can be taken as being $a = 0.2 \times 10^{-6}\ m^2/s$*, from* $10°C$ *to* $30°C$ *the change will penetrate into the body of the rock so that the temperature distribution will correspond to the curves in Fig. 14.2.*

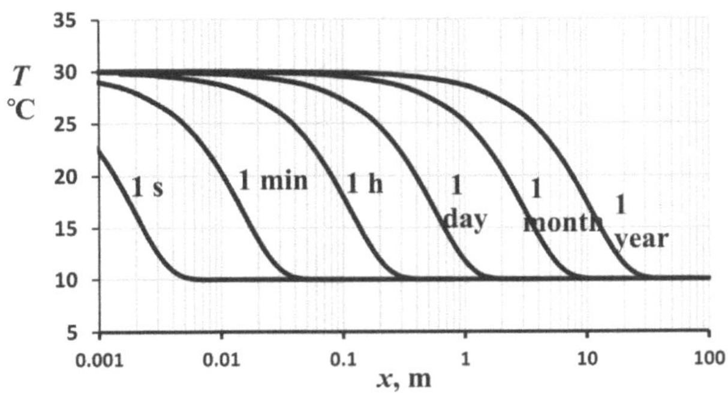

Fig. 14.2 Temperature distribution in a rock after the temperature step.

14.2.3 The second fundamental kind

This boundary conditions assume a **constant step change of the transfer intensity** at the boundaries of the system. An immediate change of the magnitude of the **gradient** $(ds/dx)_{x=0}$ occurs along with it. The condition at the fluid boundary is then

$$\left.\frac{\partial s}{\partial x}\right|_{x=0} = s'_w \quad \text{for } x = 0 \quad \text{at any time } t > 0 \tag{14.24}$$

The starting state is considered again as a fluid in rest, it is

$$\frac{\partial s}{\partial x} = 0 \text{ and } s = 0 \quad \text{at } t = 0 \quad \text{for any place } x > 0 \tag{14.25}$$

The balance equation (14.4) can be differentiated with respect to x will result in

$$\frac{\partial}{\partial t}\left(\frac{\partial s}{\partial x}\right) = B\frac{\partial^2}{\partial x^2}\left(\frac{\partial s}{\partial x}\right) \tag{14.26}$$

Obviously, the first step of the solution will be a procedure identical to the equations numbered from Eqs. (14.11)–(14.17) so that relationship (14.27) will apply

$$\frac{\partial s}{\partial x} \equiv \frac{1}{2\sqrt{Bt}} \frac{\partial s}{\partial \xi} = s'_w \operatorname{erfc}(\xi) \tag{14.27}$$

Further integration will result in

$$\begin{aligned} s &= s_0 + s'_w \frac{2\sqrt{Bt}}{\sqrt{\pi}} \psi_2(\xi) \\ &= s_0 + s'_w \frac{2\sqrt{Bt}}{\sqrt{\pi}} \left[\exp\left(-\xi^2\right) - \sqrt{\pi}\,\xi \operatorname{erfc}(\xi)\right] \end{aligned} \tag{14.28}$$

where

$$\psi_2(\xi) = \exp\left(-\xi^2\right) - \sqrt{\pi}\,\xi \operatorname{erfc}(\xi) \tag{14.29}$$

The thickness of the boundary layer defined by relationship (14.19) is given here by

$$\delta_I = \delta_{2I}(t) \equiv \frac{2}{\sqrt{\pi}} \sqrt{Bt} \approx 1.128\sqrt{Bt} \tag{14.30}$$

When using dimensionless arguments (x/δ_2), then the scalar quantity s will have the form

$$\begin{aligned} \frac{s(x,t) - s_0}{s'_w \delta_2(t)} &= \Psi_2\left(\frac{x}{\delta_2}\right) \\ &= \exp\left[\frac{-1}{\pi}\left(\frac{x}{\delta_2}\right)^2\right] - \left(\frac{x}{\delta_2}\right) \operatorname{erfc}\left[\frac{1}{\sqrt{\pi}}\left(\frac{x}{\delta_2}\right)\right] \end{aligned} \tag{14.31}$$

This relationship is also depicted in Fig. 14.2.

This function also starts at $\Psi_2(0) = 1$ and converges with growing ξ very fast to zero. For example, $\Psi_2(2) \approx 0.012$, $\Psi_2(3) \approx 0.00017$, and $\Psi_2(4) \approx 0.00000053$.

The fields of the individual quantities will be then expressed by the following relationships (14.32)–(14.34):

$$u_y(x,t) = u_0 - \frac{\tau_w \sqrt{\nu t}}{\mu} \left[\frac{2}{\sqrt{\pi}} \exp\left(-\frac{x^2}{4\nu t}\right) - \frac{x}{\sqrt{\nu t}} \operatorname{erfc}\left(\frac{x}{2\sqrt{\nu t}}\right)\right] \tag{14.32}$$

$$T(x,t) = T_0 + \frac{q_w\sqrt{at}}{k}\left[\frac{2}{\sqrt{\pi}}\exp\left(-\frac{x^2}{4\nu t}\right) - \frac{x}{\sqrt{\nu t}}\operatorname{erfc}\left(\frac{x}{2\sqrt{\nu t}}\right)\right] \quad (14.33)$$

$$C_A(x,t) = C_0 + n_w\sqrt{\frac{t}{\mathcal{D}}}\left[\frac{2}{\sqrt{\pi}}\exp\left(-\frac{x^2}{4\mathcal{D}t}\right) - \frac{x}{\sqrt{\mathcal{D}t}}\operatorname{erfc}\left(\frac{x}{2\sqrt{\mathcal{D}t}}\right)\right] \quad (14.34)$$

The concretization of the individual partial results is summarized in Table 14.3.

In reality, this second fundamental problem represents, e.g.:

- **For momentum transfer**: step change of a tangential force acting to the wall velocity.
- **For heat transfer**: it is the heating with a flat heat source (an electric heater) or the absorption of heat radiation **by a surface (sun radiation). An important instance of the second boundary condition is the case of a** zero heat flow through an interface. This is the case of an **isolated wall** for which the temperature gradient is zero. In an axisymmetric system, it can be also assigned to the axis of symmetry of the observed system.
- **For mass transfer**: it may be when the mass flow is controlled by a layer of constant resistance, e.g., a membrane.

Example:

A rock having a thermal conductivity $k = 0.5\,W/(m\,K)$ and thermal diffusivity $a = 0.2 \times 10^{-6}\,m^2/s$, will be the subject of constant heating with the intensity of the heat flow of $q_W = 500\,W/m^2$ (this being the summer sunny heat rate at noon in Central Europe). Close under the surface, the temperature gradient will be $dT/dx = 1000\,K/m$ and the temperature response will move from the original $T_\infty = 10°C$ to a temperature distribution shown by the curves in Fig. 14.3. (Note: The night cooling off due to the radiation of heat into the bright sky has approximately the same intensity. Therefore, the curve calculated for 1 day has no more any sense in the context of the sun heating.)

14.2.4 The transfer into a semispace while condition of the boundary oscillates

If the temperature of the boundary harmonically oscillates around the value T_0 with a frequency f and an amplitude ΔT_0, then the boundary condition will be

$$s = s_0 + \Delta s_0 \sin(2\pi f t) \quad \text{for } x = 0 \text{ during the whole examined time} \quad (14.35)$$

Table 14.3 Unsteady unidirectional transfer in half-space; second fundamental kind of the boundary conditions.

Quantity transferred	Momentum	Heat	Species
Initial condition for			
$t=0 \quad s=s_0$	$u_y = u_0$	$T = T_0$	$C_A = C_0$
Boundary condition, gradient of the wall	Velocity gradient,	Temperature gradient,	Concentration gradient,
for $x=0 \quad \frac{\partial s}{\partial x} = s'_w$	$u'_w \equiv \left.\frac{\partial u_y}{\partial x}\right\|_{x=0} = \frac{\tau_w}{\mu}$	$T'_w \equiv \left.\frac{\partial T}{\partial x}\right\|_{x=0} = \frac{-q_w}{k}$	$C'_w \equiv \left.\frac{\partial C_A}{\partial x}\right\|_{x=0} = \frac{-n_w}{\mathcal{D}}$
Value at the wall	Velocity	Temperature	Concentration
$\psi_2(0,t)=1$	$w_2(0,t)=1$	$\vartheta_2(0,t)=1$	$\chi_2(0,t)=1$
Dimensionless coordinate			
$\xi \equiv \frac{x}{2\sqrt{Bt}}$	$\xi \equiv \frac{x}{2\sqrt{\nu t}}$	$\xi \equiv \frac{x}{2\sqrt{at}}$	$\xi \equiv \frac{x}{2\sqrt{\mathcal{D}t}}$
Definition			
$\psi_2 \equiv \frac{s-s_0}{-s'_w}\sqrt{\frac{\pi}{4Bt}}$	$w_2 \equiv \frac{u_y-u_0}{-u'_w}\sqrt{\frac{\pi}{4\nu t}}$	$\vartheta_2 \equiv \frac{T-T_0}{-T'_w}\sqrt{\frac{\pi}{4at}}$	$\chi_2 \equiv \frac{C_A-C_0}{-C'_w}\sqrt{\frac{\pi}{4\mathcal{D}t}}$
Dimensionless profile			
$\psi_2(\xi) = \exp\left(-\xi^2\right) - \sqrt{\pi}\,\xi\,\mathrm{erfc}(\xi)$	$w_2 = \exp\left(-\xi^2\right) - \sqrt{\pi}\,\xi\,\mathrm{erfc}(\xi)$	$\vartheta_2 = \exp\left(-\xi^2\right) - \sqrt{\pi}\,\xi\,\mathrm{erfc}(\xi)$	$\chi_2 = \exp\left(-\xi^2\right) - \sqrt{\pi}\,\xi\,\mathrm{erfc}(\xi)$
Gradient at the wall	Velocity gradient,	Temperature gradient,	Concentration gradient,
$\left.\frac{\partial \psi_2}{\partial x}\right\|_{x=0} = \frac{-1}{\sqrt{\pi Bt}}$	$\left.\frac{\partial w_2}{\partial x}\right\|_{x=0} = \frac{-1}{\sqrt{\pi \nu t}}$	$\left.\frac{\partial \vartheta_2}{\partial x}\right\|_{x=0} = \frac{-1}{\sqrt{\pi at}}$	$\left.\frac{\partial \chi_2}{\partial x}\right\|_{x=0} = \frac{-1}{\sqrt{\pi \mathcal{D}t}}$
Boundary layer thickness	Momentum boundary layer	Thermal boundary layer	Concentration boundary layer
$\delta_I \equiv \frac{s_w - s_0}{-s'_w}$	$\delta_{2I}(t) \equiv \frac{2}{\sqrt{\pi}}\sqrt{\nu t}$	$\Delta_{2I} \equiv \frac{2}{\sqrt{\pi}}\sqrt{at}$	$\Delta_{2I} \equiv \frac{2}{\sqrt{\pi}}\sqrt{\mathcal{D}t}$
Value at the wall	Velocity	Temperature	Concentration
$\psi_2(0,t)=1$	$w_2(0,t)=1$	$\vartheta_2(0,t)=1$	$\chi_2(0,t)=1$
Value at the wall	Velocity	Temperature	Concentration
$s_w = s_0 - \frac{2s'_w}{\sqrt{\pi}}\sqrt{Bt}$	$u_w = u_0 - \frac{2\tau_w}{\mu\sqrt{\pi}}\sqrt{\nu t}$	$T_w = T_0 + \frac{2q_w}{k\sqrt{\pi}}\sqrt{at}$	$C_w = C_0 + \frac{2\,n_w}{\sqrt{\pi \mathcal{D}}}\sqrt{t}$

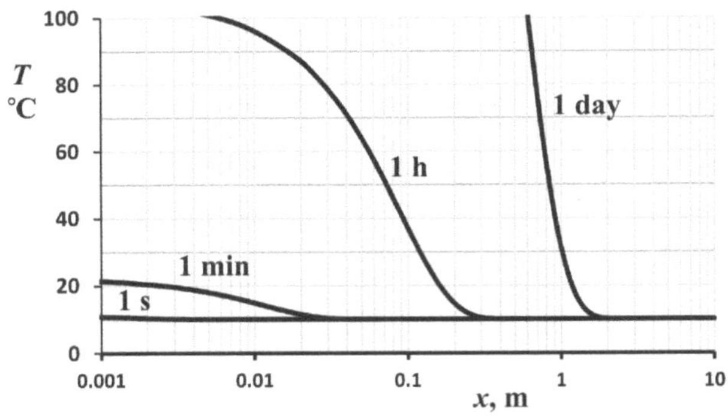

Fig. 14.3 Temperature distribution in a rock after the heating began by a constant speed.

It can be calculated that in only a few time periods the velocity field will begin to settle into a harmonic oscillation with an amplitude $\Delta s(x)$, dropping exponentially with the decreasing distance x from the boundary, while this oscillation is delayed from the oscillation of the surface (a phase shift) by $\Delta t(x)$. Therefore, it will be

$$s(x, t) = s_0 + \Delta s(x) \sin\left[2\pi f (t - \Delta t(x))\right] \tag{14.36}$$

With the exponentially deceasing local amplitude

$$\Delta s(x) = \Delta s_0 \exp\left(-x\sqrt{\frac{\pi f}{B}}\right) \tag{14.37}$$

while the phase shift grows linearly

$$\Delta t(x) = \frac{x}{2\sqrt{\pi B f}} \tag{14.38}$$

Example:

In a rock, the thermal diffusivity of which can be taken as being $a = 0.2 \times 10^{-6}\ m^2/s$, the temperature oscillates harmonically around $T_0 = 10°C$ with an amplitude $\Delta T_0 = 20K$. In the course of a year, the frequency of the oscillation is $f = 1/(365 \times 24 \times 3600) \approx 3 \times 10^{-8}\ s^{-1}$. With the help of relationships (14.37) and (14.38), it will be found that in a cellar that is 5 m deep the temperature will oscillate with an amplitude $\Delta T(5) \approx 0.66$ K and with a delay of $\Delta t(5) \approx 210$ days. In a real case, the fluctuation could be weaker due to freezing, thaw, condensation, and evaporation of the ground humidity.

Example:

In a rock, the thermal diffusivity of which can be taken as being $a = 0.2 \times 10^{-6}\ m^2/s$*, the temperature of the ground surface during a summer day oscillates around* $20°C$ *with an amplitude of* $10K$ *and a frequency of* $f = 1/(24 \times 3600) \approx 1 \times 10^{-5}\ s^{-1}$*. With the use of the relationships (14.37) and (14.38), the oscillation in the depth* $0.1\,m$ *has a frequency of* $T(0.1) \approx 2.6K$ *and a delay of* $\Delta t(0.1) \approx 5h$*.*

14.3 The integral parameters of heat conduction

So far, the distribution of temperatures was considered to be the solution of the problem of heat conduction. For a unidirectional heat conduction, this is the temperature profile $T(x)$ in the interval $x_0 \leq x \leq x_1$. In simple cases dealt with so far, the temperature profile and its formula were found to be sufficiently well arranged so that they will be kept also for the future.

In more complex cases that will be dealt with next, though the knowledge of the temperature distribution is of interest, yet in the engineering practice a few of integral parameters used to be satisfactory for the design and evaluation of transport processes.

There are several integral quantities that can be accessed by experiment. One of those is the **mean temperature** T_M. This is the temperature that a system would reach if it were left from a chosen moment for a sufficiently long time isolated. For the case of an unsteady heat conduction, it is possible to define a mean temperature of a given volume V as a function of time $T_M(t)$ so that it would characterize the instantaneous heat content of the material in a given volume.

$$T_M(t) \equiv \frac{\iiint \rho c_P T(t) \mathrm{d}V}{\iiint \rho c_P \mathrm{d}V} \tag{14.39}$$

This is the definition of the mean temperature in a stationary volume. Later a more complex definition of the so-called **mixed-mean temperature** will be introduced to characterize the temperature in a fluid flowing out through a given cross section.

In many cases, both the density and the heat capacity can be considered constant so that a simplified relationship follows:

$$T_M(t) = \frac{1}{V} \iiint T(t) \mathrm{d}V \quad \text{for } \rho c_P \approx \text{const.} \tag{14.40}$$

If the function T_M is known, then the speed of the heat Q accumulation in a given volume ΔV can be determined for a given moment by

$$\frac{\mathrm{d}Q}{\mathrm{d}t} = \rho c_P \Delta V \frac{\mathrm{d}T_M}{\mathrm{d}t} \tag{14.41}$$

If there is no internal source of heat in the system, then the rate of heat accumulation is only due to the conduction through the boundary surface ΔS_W so that

$$\frac{dQ}{dt} = q_W \Delta S_W \tag{14.42}$$

where q_W is the local intensity of the heat flow through an element of the heat-exchanging boundary surface ΔS_W.

It can be expected that the heat flow through a boundary surface (usually a wall) into the system will depend on the driving force. Usually, this is the difference between the wall temperature T_W and the mean temperature T_M. It is, therefore, customary to introduce a **heat transfer coefficient** by a definition

$$h \equiv \frac{q_W}{T_W - T_M} \tag{14.43}$$

The value of the heat transfer coefficient is also expressed by a relationship between temperatures

$$h \equiv \frac{\rho c_P}{(T_W - T_M)} \frac{\Delta V}{\Delta S_W} \frac{\mathrm{d}T_M}{\mathrm{d}t} \tag{14.44}$$

In any case, it needs to assign suitable values q_W, T_M, T_W.

It is often useful to know how much of the heat is transferred in the whole interval between the temperature jump and the final time t. Therefore, a mean coefficient of heat transfer is defined as

$$\bar{h}(z) \equiv \frac{\int_0^t q\, dt}{t\,\overline{\Delta T}} \tag{14.45}$$

whereby for the characteristic temperature difference is usually chosen the arithmetic mean

$$\overline{\Delta T} \equiv \frac{[T_W(0) - T_M(0)] + [T_W(t) - T_M(t)]}{2} \tag{14.46}$$

14.3.1 Dimensionless numbers

For heat conduction, it is customary to set up a **Fourier number**

$$Fo \equiv \frac{kt}{\rho c_P L^2} = \frac{at}{L^2} \tag{14.47}$$

which in reality is a dimensionless time coordinate.

For dimensionless form of the heat transfer coefficient, it is customary to set up an **instant Biot number**

$$Bi \equiv \frac{hL}{k} \tag{14.48}$$

It comes out that

$$Bi \equiv \frac{L^2 \rho c_P}{k(T_W - T_M)} \frac{\mathrm{d}T_M}{\mathrm{d}t} \tag{14.49}$$

By introducing the Fourier number the relationship (14.49) can simply be presented as

$$Bi = \frac{1}{4(T_W - T_M)} \frac{\mathrm{d}T_M}{\mathrm{d}Fo} \tag{14.50}$$

As long as the temperature distribution is characterized by the thermal boundary layer Δ_I, the Biot number will be then the ratio of the longitudinal variables

$$Bi = \frac{L}{\Delta_I} \tag{14.51}$$

Because there is no characteristic length dimension in the half-space, it will simply apply there

$$Bi \sqrt{Fo} = \text{const.} \tag{14.52}$$

The **mean Biot number** turns out to be

$$\overline{Bi}(t) = \frac{\overline{h}L}{k} \tag{14.53}$$

14.4 One-directional unsteady transfer in a finite space

14.4.1 The strategy of solving the equations of unsteady unidirectional heat conduction in a finite layer

A prototype exercise in heat transfer is the heat conduction in a flat layer having a thickness L in which on one side ($x=0$) a sudden jump-temperature change occurs at the moment $t=0$, while the other wall ($x=L$) remains at the initial stage. Four fundamental combinations of boundary conditions can be defined for this simplest version

First combination $T=T_W$ for $x=0$ and $dT/dx=0$ for $x=L$ (14.54)

Second combination $dT/dx=-q_W/k$ for $x=0$ and $dT/dx=0$ for $x=L$ (14.55)

Third combination $T=T_W$ for $x=0$ and $T=T_0$ for $x=L$ (14.56)

Fourth combination $dT/dx=-q_W/k$ for $x=0$ and $T=T_0$ for $x=L$ (14.57)

while the initial condition is

$$T=T_0 \quad \text{for } t=0 \tag{14.58}$$

14.4.2 The asymptote for short times $Fo \to 0$

If only a short time elapsed from the moment when the change occurred, that is, the $Fo \ll 1$, the influence of the adverse side of the plate is negligible and its boundary condition can be moved into infinity. The solution for a half space is applicable to this case.

According to Eq. (14.19), in **first and third combination** the following will apply:

$$\lim_{Fo\to 0}\Delta_1 = \sqrt{\pi}\sqrt{at} \tag{14.59}$$

and then

$$\lim_{Fo\to 0}\left(Bi\sqrt{Fo}\right) = \frac{1}{\sqrt{\pi}} \tag{14.60}$$

The temperature distribution will be

$$\lim_{Fo\to 0} T = T_0 + (T_W - T_0)\ \mathrm{erfc}\left(\frac{\sqrt{\pi}}{2}\frac{x}{\Delta_1}\right) \tag{14.61}$$

According to Eq. (14.31), in **second and fourth combination**

$$\lim_{Fo\to 0} \Delta_2 = \frac{2\sqrt{at}}{\sqrt{\pi}} \tag{14.62}$$

and

$$\lim_{Fo\to 0} T_W = T_0 + \frac{q_W}{k}\Delta_2 \tag{14.63}$$

and, therefore,

$$\lim_{Fo\to 0}\left(Bi\sqrt{Fo}\right) = \frac{\sqrt{\pi}}{2} \tag{14.64}$$

The temperature distribution will be

$$\lim_{Fo\to 0} T = T_0 + \frac{q_W}{k}\Delta_2\left\{\exp\left[-\frac{1}{\pi}\left(\frac{x}{\Delta_2}\right)^2\right] - \left(\frac{x}{\Delta_2}\right)\mathrm{erfc}\left[\frac{1}{\sqrt{\pi}}\frac{x}{\Delta_2}\right]\right\} \tag{14.65}$$

14.4.3 The steady-state asymptote *Fo*→∞

In the other case of a long duration ($Fo \to \infty$), the temperature profile becomes steady

$$T_\infty\left(\frac{x}{L}\right) \equiv \lim_{Fo\to\infty} T\left(Fo, \frac{x}{L}\right) \tag{14.66}$$

that can be derived by a procedure discussed in Chapter 6.

For **first combination**

$$T_\infty = T_{M\infty} = T_W \tag{14.67}$$

For **third combination**:

$$T_\infty = T_W + \frac{x}{L}(T_0 - T_W) \tag{14.68}$$

and

$$T_{M\infty} = \frac{T_0 + T_W}{2} \tag{14.69}$$

For **fourth combination**:

$$T_\infty \equiv T_0 + \frac{q_W L}{k}\left(1 - \frac{x}{L}\right) \tag{14.70}$$

$$\text{and } T_{M\infty} = T_0 + \frac{q_W L}{2k} \tag{14.71}$$

$$\text{and } T_{W\infty} = T_0 + \frac{q_W L}{k} \tag{14.72}$$

For **second combination**, the steady quantity is only the heat flow

$$\lim_{Fo\to\infty} q\left(\frac{x}{L}\right) = q_W\left(1 - \frac{x}{L}\right) \tag{14.73}$$

By the integration of this relationship, the temperature distribution will be obtained that moves uniformly with time

$$\lim_{Fo\to\infty} T\left(Fo, \frac{x}{L}\right) = T_0 + \frac{q_W L}{k}\left[Fo - \frac{x}{L} + \frac{1}{2}\left(\frac{x}{L}\right)^2 + \frac{1}{3}\right] \tag{14.74}$$

and

$$\lim_{Fo\to\infty} T_W = T_0 + \frac{q_W L}{k}\left[Fo + \frac{1}{3}\right] \tag{14.75}$$

while it holds in the whole range

$$T_M = T_0 + \frac{q_W t}{\rho c_P L} \tag{14.76}$$

14.4.4 A complete solution $0 < Fo \leq \infty$

There are more complex calculations leading to solutions valid for the full range of Fourier numbers. Apart from the numerical solution, there is also an analytical solution using the **method of separation of variables** that can unfold the solution of partial differential equations in a solution of an infinite converging series of ordinary differential equations which are somewhat easier to solve.

The process of the separation of variables is in this case based on the expression of the dependence of the dimensionless temperature into a couple of dimensionless variables (x/L) and Fo by the expansion in an infinite series

$$\frac{T - T_\infty}{T_0 - T_\infty} = \sum_{n=1}^{\infty} A_n F_n(x/L)\,\exp\left(-b_n^2\,Fo\right) \tag{14.77}$$

Its derivation and further details of the determination of the so-called eigenfunctions $F_n(x/L)$, the eigenvalues b_1^2 and of the multipliers A_n are given in Section 21.2.

14.4.5 The improved asymptote for long duration $Fo \gg 1$

This series (14.76) converges very fast for large $Fo \gg 1$. For the determination of the temperature field just the lowest eigenvalue b_1^2 and the single function $F_1(x/L)$ will suffice

$$\lim_{Fo\to\infty}\frac{T-T_\infty}{T_0-T_\infty}=A_1F_1\left(\frac{x}{L}\right)\exp\left(-b_1^2Fo\right) \tag{14.78}$$

This relationships describe how the temperature field exponentially converges to a steady-state T. Already for $Fo>0.1$ it is a very good approximation. Inversely, in the case of a small Fo, a series with a large number of terms would be needed. The circumstance that the eigenfunction comes out as a simple goniometric functions is the specialty of the solution of heat transfer in the simplest geometry, in a flat layer. The knowledge of the Bessel functions is needed to solve the heat conduction in a cylindrical rod. To solve the heat conduction in even more complex geometries functions are needed that are not currently tabulated.

It will apply in specific case of the boundary conditions in a finite layer For **first combination**

$$\lim_{Fo\to\infty}\frac{T-T_0}{T_w-T_0}=1-\frac{4}{\pi}\sin\left(\frac{\pi x}{2L}\right)\exp\left(-\frac{\pi^2}{4}Fo\right) \tag{14.79}$$

and

$$\lim_{Fo\to\infty}\frac{T_M-T_0}{T_w-T_0}=1-2\exp\left(-\frac{\pi^2}{4}Fo\right) \tag{14.80}$$

and the Biot number

$$\lim_{Fo\to\infty}Bi=\frac{\pi^2}{4}\approx 2.467\ldots \tag{14.81}$$

For **second combination**

$$\lim_{Fo\to\infty}\frac{k(T-T_0)}{q_wL}=\frac{1}{2}\left(\frac{x}{L}\right)^2-\frac{x}{L}+\frac{1}{3}+Fo-\frac{2}{\pi^2}\cos\left(\pi\frac{x}{L}\right)\exp\left(-\pi^2Fo\right) \tag{14.82}$$

and

$$\lim_{Fo\to\infty}\frac{k(T_M - T_0)}{q_w L} = Fo \tag{14.83}$$

and the Biot number

$$\lim_{Fo\to\infty} Bi = 3 \tag{14.84}$$

For **third combination**

$$\lim_{Fo\to\infty}\frac{T - T_0}{T_w - T_0} = 1 - \frac{x}{L} - \frac{2}{\pi}\sin\left(\pi\frac{x}{L}\right)\exp\left(-\pi^2 Fo\right) \tag{14.85}$$

and

$$\lim_{Fo\to\infty}\frac{T_M - T_0}{T_w - T_0} = \frac{1}{2} - \frac{4}{\pi^2}\exp\left(-\pi^2 Fo\right) \tag{14.86}$$

and the Biot number

$$\begin{aligned}\lim_{Fo\to\infty} Bi &= 2 + \left(4 - \frac{16}{\pi^2}\right)\exp\left(-\pi^2 Fo\right)\\ &\approx 2 + 2.379\exp(-9.87 Fo)\end{aligned} \tag{14.87}$$

For **fourth combination**

$$\lim_{Fo\to\infty}\frac{k(T - T_0)}{q_w L} = 1 - \frac{x}{L} - \frac{8}{\pi^2}\cos\left(\frac{\pi}{2}\frac{x}{L}\right)\exp\left(\frac{-\pi^2}{4} Fo\right) \tag{14.88}$$

and

$$\lim_{Fo\to\infty}\frac{k(T_M - T_0)}{q_w L} = \frac{1}{2} - \frac{16}{\pi^3}\exp\left(\frac{-\pi^2}{4} Fo\right) \tag{14.89}$$

and the Biot number

$$\begin{aligned}\lim_{Fo\to\infty} Bi &= 2 + \frac{32}{\pi^2}\left(1 - \frac{2}{\pi}\right)\exp\left(\frac{-\pi^2}{4} Fo\right)\\ &\approx 2 + 1.178\exp(-2.467 Fo)\end{aligned} \tag{14.90}$$

How well the individual asymptotes fit is shown in Table 14.4.

Table 14.4 The usability of the individual approximations in which the error does not exceed 5%.

Combination	$Fo \to 0$	$Fo \gg 1$	$Fo \to \infty$
1	$Fo < 0.13$	$Fo > 0.05$	$Fo > 1.1$
2	$Fo < 0.50$	$Fo > 0.02$	$Fo > 0.22$
3	$Fo < 0.13$	$Fo > 0.05$	$Fo > 0.33$
4	$Fo < 0.50$	$Fo > 0.03$	$Fo > 1.2$

Complete solutions of temperature profiles in a flat plate are presented graphically for the selected boundary conditions in Fig. 14.4.

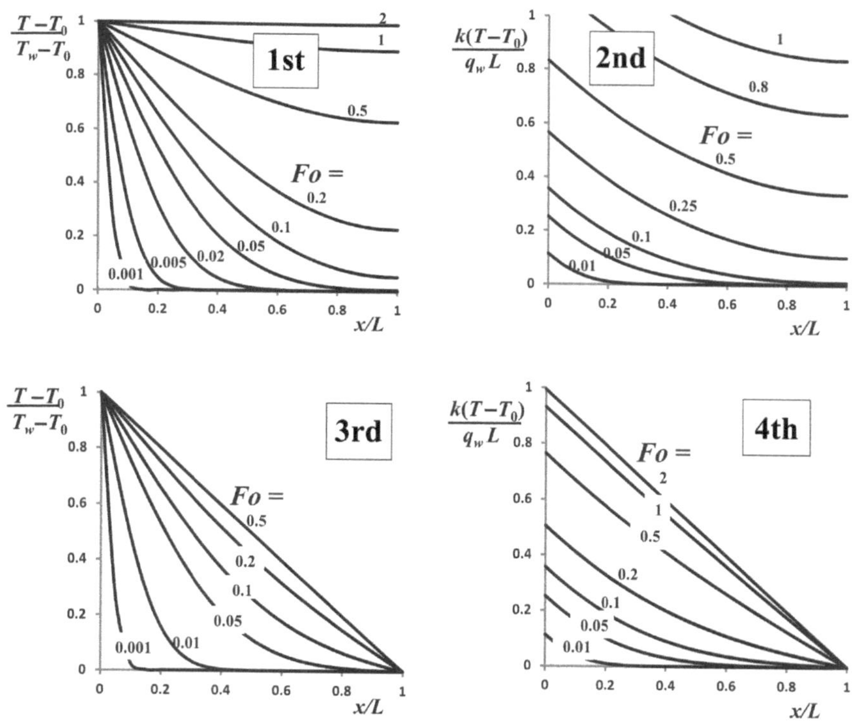

Fig. 14.4 The normalized temperature fields in a flat layer for the fundamental combinations. The parameter of the curves is the Fourier number.

How well the asymptotes fit the rate of the heat transfer can be also deduced from the relationship $Bi(Fo)$ shown in Fig. 14.5.

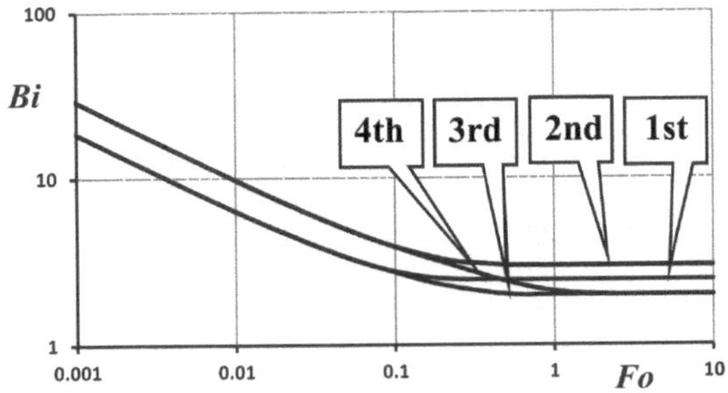

Fig. 14.5 The Biot number in a flat layer for the fundamental combinations.

14.5 Summary

One of the few flows depending on two coordinates that can be solved by simple partial differential Navier-Stokes equations is the unidirectional unsteady flow. The solution is mathematically identical to the solution of the unsteady heat conduction or diffusion (the second Fick's law).

To solve the equations of the unidirectional transfer, the setting of the boundary conditions is essential. The basic task is the examination of the development of the velocity, temperature, or concentration field starting from the initial constant state. The transfer is started by a jump change; either of the value at the boundary or of its derivative that is the jump of momentum, heat, or mass flow through the boundary.

If the examined space can be considered to be unlimited half-space, then the problem can be solved by a method of combining the variables. A new variable can be adopted that combines the influence of the distance and time elapsed. With this new variable, it is possible to transform the partial differential equation to an ordinary differential one. Then, the characteristic thickness of the boundary layer becomes directly proportional to the square root of the product of the kinematic transport coefficient (i.e., kinematic viscosity, thermal diffusivity, diffusion coefficient, resp.) and the time elapsed.

This task has a simple solution even for the case of oscillating boundary conditions.

Usually, the solution of the problem does not have to have the format of the complete knowledge of the temperature field. It is quite satisfactory to know the integral quantities that present the main balances. Suitable information is the mean temperature and the mean heat flow. It is customary to introduce a coefficient of heat transfer or to express it in the form of the Biot number.

In a way, the Biot number expresses the dimensionless temperature gradient at the boundary. Thus, it could be used for the determination of the thickness of the thermal boundary layer, where the main portion of the heat transfer occurs.

From here it is possible to determine when, in the case of heat conduction in a finite space, the temperate field could be interpreted, for short time periods, as if this were the transfer into a free space.

Otherwise, a solution of partial differential equations is needed. The result in this case can be expressed by a series of geometric functions. The convergence of this series for higher Fourier numbers is very fast.

CHAPTER FIFTEEN

The flow in two dimensions

Contents

15.1 The stream function

For some steady flows, it is possible by using a suitable transformation of the coordinate system, to achieve a setup when the nonzero velocity components exist only in two directions. This is demonstrated in the Cartesian coordinate system for a velocity field having only two components $u_x(x, y)$ and $u_y(x, y)$.

Transport and Surface Phenomena
https://doi.org/10.1016/B978-0-12-818994-8.00015-4

In this case, it is possible to find a field of a scalar quantity $\psi(x,y)$ for which it holds that

$$\frac{\partial \psi}{\partial y} = u_x \quad \frac{\partial \psi}{\partial x} = -u_y \tag{15.1}$$

This function called the **stream function** satisfies the equation of continuity of an incompressible fluid because it holds that

$$\frac{\partial u_x}{\partial x} + \frac{\partial u_y}{\partial y} = \frac{\partial^2 \psi}{\partial x\,\partial y} - \frac{\partial^2 \psi}{\partial y\,\partial x} = 0 \tag{15.2}$$

It further holds that no fluid flows through the planes for which $\psi =$ const. They represent the **streamlines**. The difference between the values of ψ is proportional to the flow rates between the relevant streamlines.

$$\int_{y_1}^{y_2} u_x dy = \int_{y_1}^{y_2} \frac{\partial \psi}{\partial y} dy = \psi_2 - \psi_1 \tag{15.3}$$

After substituting ψ into the two-directional Navier-Stokes equation for $\boldsymbol{x}$ direction one gets

$$\rho\left(\frac{\partial^2 \psi}{\partial t\,\partial y} + \frac{\partial \psi}{\partial y}\frac{\partial^2 \psi}{\partial x\,\partial y} - \frac{\partial \psi}{\partial x}\frac{\partial^2 \psi}{\partial y^2}\right) = \rho g_x - \frac{\partial p}{\partial x} + \mu\left(\frac{\partial^3 \psi}{\partial x^2 \partial y} + \frac{\partial^3 \psi}{\partial y^3}\right) \tag{15.4}$$

and for the y direction

$$\rho\left(-\frac{\partial^2 \psi}{\partial t\,\partial x} - \frac{\partial \psi}{\partial y}\frac{\partial^2 \psi}{\partial x^2} + \frac{\partial \psi}{\partial x}\frac{\partial^2 \psi}{\partial x\,\partial y}\right) = \rho g_y - \frac{\partial p}{\partial y} + \mu\left(-\frac{\partial^3 \psi}{\partial x^3} - \frac{\partial^3 \psi}{\partial x\,\partial y^2}\right) \tag{15.5}$$

The driving force can be eliminated from the above two equations because it holds

$$\frac{\partial}{\partial y}\left(\rho g_x - \frac{\partial p}{\partial x}\right) = \frac{\partial}{\partial x}\left(\rho g_y - \frac{\partial p}{\partial y}\right) \tag{15.6}$$

so that the relevant terms fall out when the first equation is differentiated by y and when the second equation, after having been differentiated by x, is taken away from the first one. In addition, usually the flow is required to be irrotational (the tensor $\boldsymbol{\Omega} = \mathbf{0}$). So for the situation described above it holds

$$\frac{\partial u_x}{\partial y} - \frac{\partial u_y}{\partial x} = 0 \quad \text{and then} \quad \frac{\partial^2 \psi}{\partial y^2} + \frac{\partial^2 \psi}{\partial x^2} = 0 \tag{15.7}$$

As a result, the two Navier-Stokes equations become one,

$$\frac{\partial(\nabla^2 \psi)}{\partial t} + \frac{\partial \psi}{\partial x}\frac{\partial(\nabla^2 \psi)}{\partial y} - \frac{\partial \psi}{\partial y}\frac{\partial(\nabla^2 \psi)}{\partial y} = \frac{\mu}{\rho}\nabla^4 \psi \tag{15.8}$$

which means that the mathematical treatment will get easier.

15.1.1 Steady two-directional creeping flow

For *Re* approaching zero, Eq. (16.8) degenerates to

$$\nabla^4 \psi = 0 \tag{15.9}$$

15.1.2 Steady two-dimensional inviscid flow

For flows having the Reynolds number approaching infinity the Euler equation for steady two-dimensional flow (16.8) degrades to

$$\frac{\partial \psi}{\partial x}\frac{\partial(\nabla^2 \psi)}{\partial y} - \frac{\partial \psi}{\partial y}\frac{\partial(\nabla^2 \psi)}{\partial y} = 0 \tag{15.10}$$

An irrotational flow can also be characterized as a potential flow in which a potential function φ exists for which

$$\frac{\partial \varphi}{\partial x} = -u_x \; \frac{\partial \varphi}{\partial y} = -u_y \tag{15.11}$$

while

$$\nabla^2 \varphi = 0 \tag{15.12}$$

Finding out the spatial distribution of the functions ψ and φ that meet the boundary conditions is a mathematical undertaking that makes use of the theory of complex variables and conform representation. The boundary conditions are, in this case, the lines of constant ψ function that copy the areas of a given speed while the lines of a constant φ function copy the given pressure potential. In nontrivial cases this subject falls beyond the scope of this text.

15.2 Combination of inviscid and creeping flow

15.2.1 Paradox of limiting cases

The Euler equation for the flow of an ideal fluid presented a challenge for the mathematicians but the addition of viscosity in the middle of the 19th century (Navier-Stokes equation) was an additional complication. An example of a successful nontrivial solution is the flow past a sphere in the two limiting cases ($Re \rightarrow 0$, $Re \rightarrow \infty$). Both these two conclusions seem to indicate that the solutions of all tasks will be some kind of interpolation between these two extremes (Fig. 15.1).

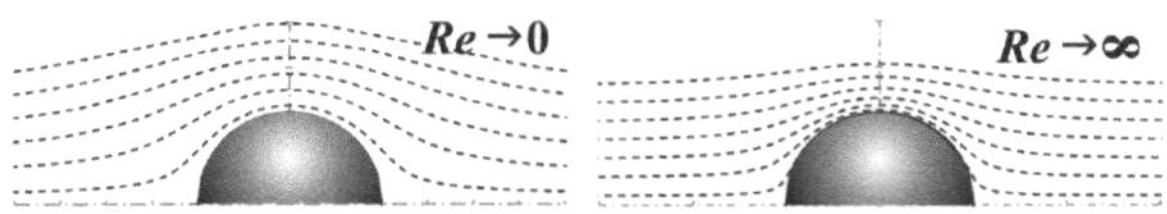

Fig. 15.1 Streamlines in a flow field past a sphere:
- Creeping flow at $Re \rightarrow 0$
- Inviscid flow at $Re \rightarrow \infty$.

The relevant experiments have shown, however, that the real picture of the flow looks differently. The reason for this phenomenon can be found in the pressure field. The distribution of pressure on the surface of the sphere is governed mainly by the Bernoulli equation. When neglecting the influence of viscosity, the Bernoulli equation predicts that the highest pressure, equal to $\frac{1}{2}\rho U^2$, will be in the incidence point (**the front critical point**) where the speed is zero. An equal value will be found in the **rear critical point**. If individual points on the surface of the sphere are characterized by angle Θ from the front critical point then at $\Theta = 90$ degrees, where the velocity is the highest, the pressure will drop to a minimum negative value as shown in Fig. 15.2.

According to the equation of the creeping flow, the pressure on the surface of a sphere drops continuously. This explains why the streamlines on the surface of a sphere are asymmetric for all the current finite values of the Reynolds number. Mathematical modeling of the hydrodynamics based purely on the assumption of either the inviscid flow or of the creeping flow fails completely.

Experiments performed toward the end of the 19th century such as those carried out by Osborn Reynolds on the nature of laminar and turbulent flow indicated that there was a need for a new approach. It was the aerodynamics

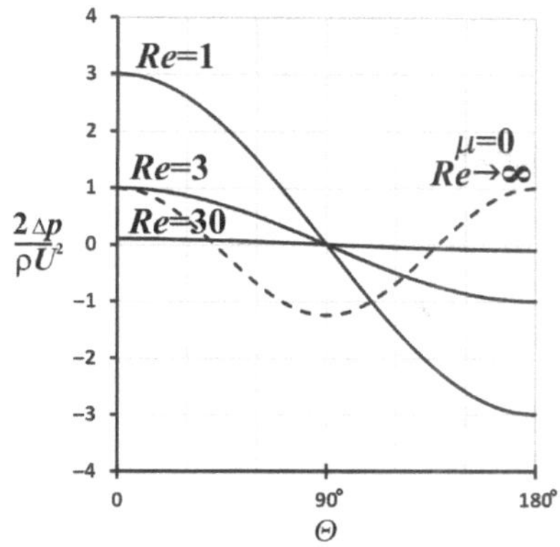

Fig. 15.2 Distribution of the normalized pressure $\Delta p/(\frac{1}{2}\rho U^2)$ at the sphere surface according to:

- the Euler equation of an inviscid potential flow
- the equation of the creeping flow.

that became the fundamental challenge, particularly after an aircraft heavier than the air rose in the air (the Wright brothers 1903). At about the same time, a new method for solving the Navier-Stokes equation appeared that was of key importance. Ludwig Prandtl (1904) called it "the flow with little influence of friction." This statement was based on an empirical observation that for bodies that are bypassed by a flowing fluid, the friction is limited to a very thin layer close to the wall. This layer is called the "boundary layer."

15.3 The flow past bodies: The boundary layer

15.3.1 The advantage of using the concept of a boundary layer: The flow past a plate—The Prandtl equation

The fact that the boundary layer is thin allows to greatly simplify the Navier-Stokes equations since it makes it possible to transform them to a format that is easier to solve. They are called the **Prandtl equations of the boundary layer**.

The nature of the boundary layer is demonstrated by the flow past a flat plate (the Prandtl problem). It turns out that the plate influences the flow only in its close vicinity (Fig. 15.3). This flow is considered to be a two-dimensional flow for which x is the coordinate along the direction of flow measured from the inlet and y is the coordinate at right angle to the wall measured from the wall. The inflowing fluid has a constant speed with only one nonzero component $u_x = U$.

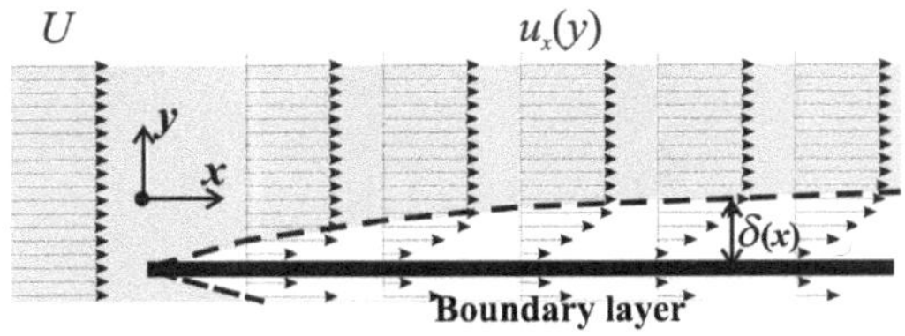

Fig. 15.3 Flow past a flat plate.

The task formulated in this way will find its application also in aerodynamics if the apparent observer moves along with the wing or with the blade of the impeller at a speed U in otherwise motionless medium.

The measurement of the velocity field close to the flat plate has shown that, under certain circumstances, there is an imaginary plane $y=\delta(x)$ at which the flow is virtually not affected by the wall so

$$u_x=U \quad \text{for} \quad y>\delta \tag{15.13}$$

while at the wall the velocity is zero. Further, a simple function f will be introduced that approximates relatively well the velocity profile at the wall

$$\frac{u_x(y,x)}{U(x)}\approx f\left(\frac{y}{\delta(x)}\right) \quad \text{pro} \quad 0\le\frac{y}{\delta}\le 1 \tag{15.14}$$

In the interval specified by (15.14) the function f grows while acquiring values between zero and one.

As long as the boundary layer is thin, $\delta \ll x$, it is realistic to expect (even for complex flows) that

$$u_y \ll u_x; \quad \frac{\partial u_x}{\partial x}\ll\frac{\partial u_x}{\partial y}; \quad \text{and} \quad \frac{\partial^2 u_x}{\partial x^2}\ll\frac{\partial^2 u_x}{\partial y^2} \tag{15.15}$$

the Navier-Stokes equation can be simplified, without a serious loss of accuracy, by leaving out the small terms

$$\frac{\partial u_x}{\partial t}+u_x\frac{\partial u_x}{\partial x}+u_y\frac{\partial u_x}{\partial y}=g_x-\frac{1}{\rho}\frac{\partial p}{\partial x}+\frac{\mu}{\rho}\frac{\partial^2 u_x}{\partial y^2} \tag{15.16}$$

This equation is known as the **Prandtl equation of boundary layer.**

From this equation, it is possible to further eliminate u_y because it follows from the equation of continuity

$$u_y=-\int_0^y\frac{\partial y_x}{\partial x}dy \tag{15.17}$$

It is known from the Bernoulli equation that the outside pressure is

$$g_x - \frac{1}{\rho}\frac{\partial p}{\partial x} = U\frac{dU}{dx} \tag{15.18}$$

The same pressure is expected to exist also inside the boundary layer so the Prandtl equation in terms of speed will be

$$\frac{\partial u_x}{\partial t} + u_x\frac{\partial u_x}{\partial x} - \frac{\partial u_x}{\partial y}\int_0^Y \frac{\partial y_x}{\partial x}dy = U\frac{dU}{dx} + \frac{\mu}{\rho}\frac{\partial^2 u_x}{\partial y^2} \tag{15.19}$$

Further, for a steady flow the time derivative disappears so the only three derivatives remaining are

$$\frac{\partial u_x}{\partial x}, \quad \frac{\partial u_x}{\partial y}, \quad \text{and} \quad \frac{\partial^2 u_x}{\partial y^2} \tag{15.20}$$

The partial differential equation of two independent variables is called in mathematics **parabolic** to distinguish them from the elliptic equations in which the second derivative with respect to x applies too. The parabolic equation can be solved either analytically or numerically stepwise from the initial condition.

So long as it is possible to neglect the second derivative in the longitudinal direction x, there is no transfer of momentum by viscous forces so whatever is going on along the x coordinate is not influenced backwards by whatever goes on at higher values of x. There is an important consequence in that the solution of Prandtl equation is equally usable for flows past walls of finite length.

If both the first and the second derivative by y occur, then two conditions that are needed are as follows:

$$\begin{aligned} u_x &= 0 && \text{for } y = 0 \\ u_x &= U(x) && \text{for } y = \delta \end{aligned} \tag{15.21}$$

When considering the dependence on x, only one condition (initial) will suffice. Then

$$u_x = U(0) \quad \text{for} \quad x = 0$$

so the zero position will be placed in the front edge of the plate.

15.3.2 The accurate solution of the Prandtl equation for the flow past a plate: The method of combining variables

The key task that Prandtl focused on first was the flow past a flat plate. If this flow is steady and if the surrounding flow has a constant value U then the equation assumes a simplified form

$$u_x \frac{\partial u_x}{\partial x} - \frac{\partial u_x}{\partial y} \int_0^y \frac{\partial u_x}{\partial x} dy = \frac{\mu}{\rho} \frac{\partial^2 u_x}{\partial y^2} \tag{15.22}$$

After Prandtl proposed approximate solution, which will be dealt with later, one of his students, Blasius suggested an elegant accurate solution.

An estimate of the order of magnitude of the members on the left-hand side of the equation gives

$$u_x \frac{\partial u_x}{\partial x} - \frac{\partial u_x}{\partial y} \int_0^y \frac{\partial u_x}{\partial x} dy = \mathrm{o}\left(\frac{U^2}{x}\right) \tag{15.23}$$

while the order of magnitude of the terms on the right-hand side will be

$$\frac{\mu}{\rho} \frac{\partial^2 u_x}{\partial y^2} = \mathrm{o}\left(\frac{\mu}{\rho} \frac{U}{\delta^2}\right) \tag{15.24}$$

The estimate of the thickness of the boundary layer follows from the parity:

$$\delta = \mathrm{o}\left(\frac{\mu x}{\rho U}\right)^{1/2} \tag{15.25}$$

Because it is likely that the velocity profiles $u_x(y/\delta)$ in the boundary layer will be mutually similar, it is advantageous to introduce instead of x and y a new combined variable that better captures the ratio y/δ

$$\xi \equiv \frac{y}{x}\left(\frac{xU\rho}{\mu}\right)^{1/2} \tag{15.26}$$

The expression in the brackets has the structure of a Reynolds number in which the distance from the leading edge is taken as the characteristic length; it is known as the **local Reynolds number** Re_x.

$$Re_x \equiv \frac{xU\rho}{\mu} \tag{15.27}$$

Because of the newly defined variable theta the trio of the boundary conditions gets reduced to a pair:

$$u_x = 0 \quad \text{for} \quad \xi = 0 \tag{15.28}$$

$$u_x = U \quad \text{for} \quad \xi = \infty \tag{15.29}$$

so a single function

$$\frac{u_x}{U} = F(\xi) \tag{15.30}$$

will suffice to reach the solution.

Because the original boundary conditions were located at zero and at an infinite position, the partial differential equation can be modified to an ordinary differential function by the method of the combination of variables. Although this ordinary differential equation cannot be solved by methods of mathematical analysis, its solution can be obtained numerically. The function is depicted in Fig. 15.4.

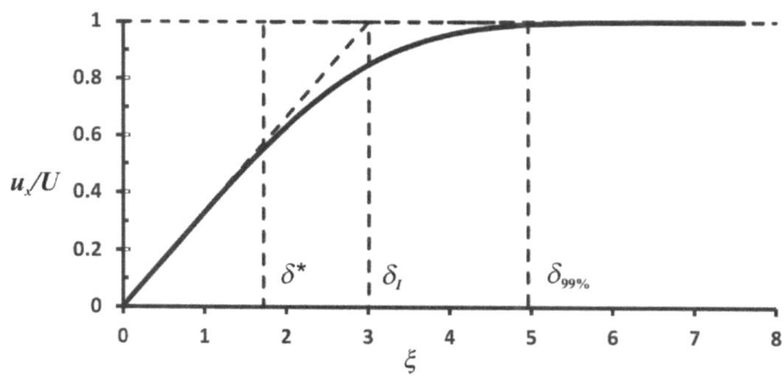

Fig. 15.4 Velocity profile and boundary layer thickness.

Which quantity should be marked as the thickness of the boundary layer is not quite unambiguous. There are three options. It was suggested to choose a thickness at which the local velocity does not differ from U by less than 1%. This option was originally suggested for the interpretation of experimental data. Here

$$\delta_{99\%} = 4.96 x Re_x^{-1/2} \tag{15.31}$$

This quantity, however, does not say much about the velocity profile inside the boundary layer. Much more objective quantity is the **displacement thickness**, δ^*, showing how the velocity profile, while maintaining the overall flow, would get divided in an immobile layer close to the wall and a region beyond this limit is not affected by the wall.

$$\delta^* \equiv \int_0^\infty \left(1 - \frac{u_x}{U}\right) dy \tag{15.32}$$

In this specific case, the above expression becomes

$$\delta^* = 1.72\, x Re_x^{-1/2} \tag{15.33}$$

From a practical point of view, the most interesting is the **thickness of the Nernst type**, δ_I, derived from the velocity gradient on the wall

$$\delta_I \equiv \frac{U}{\left.\frac{\partial u_x}{\partial y}\right|_{y=0}} \tag{15.34}$$

W.H. Nernst (1864–1941) was a German physical chemist who introduced such a concept of a boundary layer to solve diffusion processes.

In the specific case dealt with in this section, the velocity gradient on the wall, and consequently also the shear rate γ, is equal to

$$\gamma \equiv \left.\frac{\partial u_x}{\partial y}\right|_{y=0} = 0.332 \frac{U}{x} Re_x^{1/2} \tag{15.35}$$

and, therefore,

$$\delta_I = 3.012 x Re_x^{-1/2} \tag{15.36}$$

The particular component of the tangential stress can be obtained by multiplying the thickness with the dynamic viscosity. It is customary (see Eq. 12.9) to express the tangential stress in a dimensionless form called the **coefficient of friction** (**friction factor**) f, that is,

$$f_x \equiv \frac{\tau_{xy}(x)}{\frac{\rho}{2} U^2} = \frac{2}{Re_x} \frac{x}{\delta_I} = 0.664 Re_x^{-1/2} \tag{15.37}$$

Usually, the friction factor is combined with the Reynolds number to form a dimensionless group

$$\frac{f_x}{2} Re_x = \frac{x}{\delta_I} \tag{15.38}$$

Useful is also the **average value of the friction factor** from the fluid entering edge of the plate

$$f \equiv \frac{\int_0^x \tau_{xy} dx}{\frac{\rho}{2} U^2 x} = \frac{\int_0^x f_x dx}{x} = 2f_x = 1.328 Re_x^{-1/2} \tag{15.39}$$

15.3.3 An approximate solution using the method of integral balances

This technique was used for the first time by Ludwig Prandtl: all the members of the equation of the boundary layer can be integrated along the y variable. It is not necessary, however, to integrate through the whole interval up to infinity. It will suffice to carry out the integration up to some value δ (this being an estimate of the boundary layer) at which it can be estimated that the velocity U is no longer affected by the wall. By the use of the equation of continuity and a bunch of mathematical manipulations (such as the integration by parts) it is possible to reach an integral balance of momentum in the x direction

$$\int_0^\delta \frac{\partial u_x}{\partial t} dy + \frac{d}{dx}\int_0^\delta u_x^2\, dy - U\frac{d}{dx}\int_0^\delta u_x dy = g_x - \frac{\delta}{\rho}\frac{\partial p}{\partial x} - \frac{\mu}{\rho}\frac{\partial u_x}{\partial y}\bigg|_{y=0} \tag{15.40}$$

To use this equation, it is necessary to first make an estimate of a likely velocity profile in the boundary layer

$$\frac{u_x}{U} = F(\eta) \quad \text{where} \quad \eta \equiv \frac{y}{\delta} \tag{15.41}$$

and a function $\delta(x)$ that ensures the validity of the momentum balance. For a steady flow past a plate the substitution into the integral balance yields

$$\delta\frac{d\delta}{dx}\int_0^1 \left(F - F^2\right) d\eta = \frac{\mu}{\rho}\frac{dF}{d\eta}\bigg|_{\eta=0} \tag{15.42}$$

Its integration with the condition

$$\delta = 0 \quad \text{for} \quad x = 0 \tag{15.43}$$

will yield for a chosen function $F(\eta)$ the sought after function $\delta(x)$ as

$$\delta = \frac{x}{Re_x^{1/2}}\sqrt{\frac{2\left.\dfrac{dF}{d\eta}\right|_{\eta=0}}{\int_0^1 \left(F - F^2\right) d\eta}} \tag{15.44}$$

Another more interesting format of this result is

$$\delta_I = \frac{x}{Re_x^{1/2}} \sqrt{\frac{2}{\left.\frac{dF}{d\eta}\right|_{\eta=0} \int_0^1 \left(F - F^2\right) d\eta}} \qquad (15.45)$$

being connected to the friction coefficient f_x.

In order to quantify the dimensionless expression under the square root, the function $F(\eta)$ must be specified. It is not without interest that it is sufficient to choose an arbitrary function that grows in the interval from zero to one and the first derivative of which, if possible, drops in the interval from a constant to zero. In this way, a different value of $\delta(x)$ will be obtained for the transition from the boundary layer to the region of the outside flow. Surprisingly, however, the values of δ_I thus obtained do not differ substantially from the exact values. As illustrated in Table 15.1, they do not differ even for a very rough estimate of $F(y/\delta) = y/\delta$. A sinusoid fit the exact solution quite well (Fig. 15.5).

Table 15.1 Estimate of the boundary layer thickness for the flow past a plate.

$F(y/\delta)$	$\delta_I/(xRe_x^{-1/2})$
Exact (Blasius)	3.012
y/δ	3.464
$2(y/\delta) - (y/\delta)^2$	2.739
$3/2(y/\delta) - 1/2(y/\delta)^3$	3.094
$4/3(y/\delta) - 1/3(y/\delta)^4$	3.236
$\sin(\pi/2 y/\delta)$	3.053

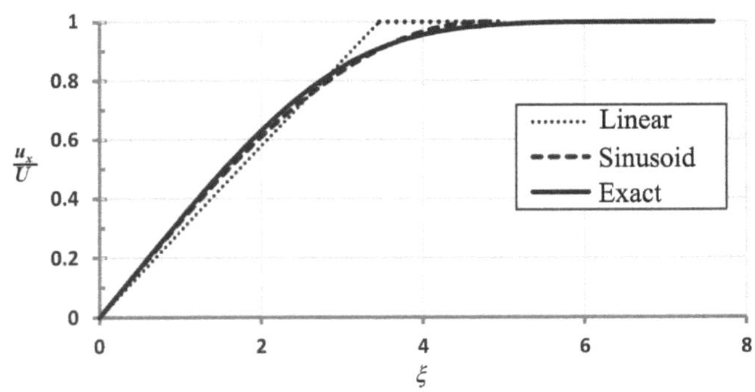

Fig. 15.5 The exact and approximate velocity profiles for a flow along a flat plate.

15.3.4 An experimental verification

Measured velocity fields in the vicinity of a bypassed plate and measured tangential stresses acting on such a plate have proved that the derived relationships describe well the reality but only for short plates when the Reynolds number does not exceed 100,000 (see Fig. 15.6). For longer plates, the validity is limited only to the initial section of the plate. As soon as the Reynolds number exceeds 100,000, the disturbances begin to appear.

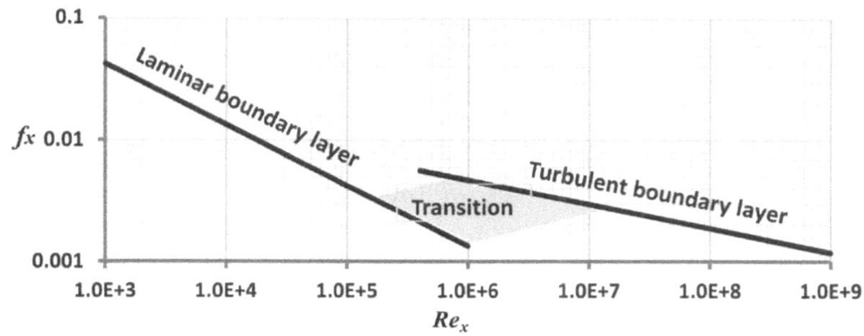

Fig. 15.6 Friction factor f_x at higher Reynolds numbers for a flow along a flat plate.

They become more evident with plates that have a rough surface or if there is an outside disturbance (such as a vibration of the equipment). As a result, a widening of the boundary layer by a turbulent zone appears in which whirling of the fluid occurs. The space can then be divided into **outside potential flow**, **turbulent boundary layer**, and **laminar sublayer** in the close vicinity of the wall. The risk of an instability occurring may be predicted from the Navier-Stokes equation according to which an accidental disturbance will be dampened by viscous forces up to a certain values of the Reynolds number. Beyond that point, there is a lack of any simple theories. Most of the knowledge about the turbulent flow is based exclusively on the experimental experience and its generalization is limited. Currently available software "solvers" have relevant hydrodynamic and aerodynamic experience already built in the expert systems. They make it possible to estimate, relatively reliably, the behavior of flowing liquids provided that the situation to be solved is not totally unusual for which there is no related, already solved case.

The flow past a flat plate is one of the basic flows for which it is known that above the **critical Reynolds number** the influence of turbulence becomes substantial. The transition can be expected to occur somewhere between Re_x equal to 100,000 and 500,000. A number of approximate empirical formulas exist for the expression of flow resistance at higher Reynolds numbers. Here is one of the most simple

$$f_x \equiv \frac{\tau_{xy}}{\frac{1}{2}\rho U^2} \approx 0.074 Re_x^{-1/5} \quad \text{for} \quad 5 \times 10^5 < Re_x < 10^7 \tag{15.46}$$

There are not more experimental data for even higher Reynolds numbers because it is difficult to reach high Reynolds numbers with liquids and any experiments with gases can be performed only in most powerful wind tunnels. In this region an additional complication appears and that is the compressibility of gases. Particularly at speeds close to the speed of sound.

15.4 The generalization of the boundary layer concept

If it is possible to assume that the boundary layer in a particular case will be thin then the distribution of the velocities and pressures outside the boundary layer could be obtained by solving the Euler equation. Subsequently, the layer, governed by the Prandtl equation, will be inserted into the boundary. Then its thickness and the distribution of the tangential forces could be figured out. In more general cases than the flow past a flat plate, the velocity U at the edge of the boundary layer is not a constant so it is first necessary to find the function $U(x)$ by solving the Euler equations. The information about the distribution of the pressure $p(x)$ on the edge of the boundary layer can be obtained in a similar way while assuming that the pressure across the boundary layer does not change further. An example of the solution of such a task will follow.

15.4.1 The flow close to the front stagnation point

One of the interesting flow situations is the impact of a stream of fluid at right angle on a flat plate (the **Hiemenz problem**)—see Fig. 15.7.

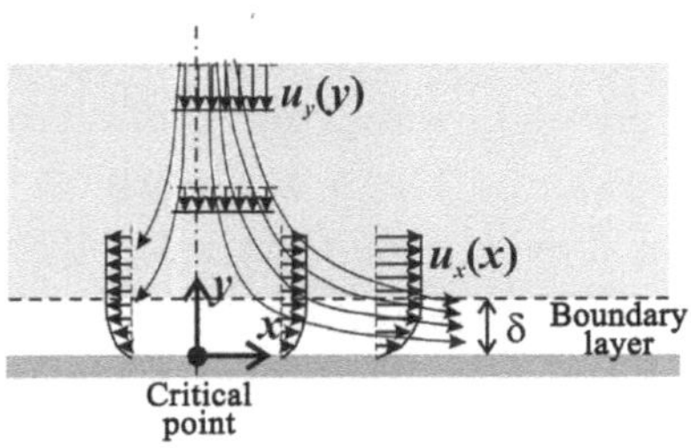

Fig. 15.7 Flow and the boundary layer close to the front stagnation point.

The simplified task to solve is the steady flow not dependent on the coordinate z. The impacting stream hitting the wall will be split at the stagnation (critical) point into two mirror-like symmetric branches. One boundary streamline is formed by a broken line formed by the axis of symmetry and the particular section of the wall. The solution of the Euler equation shows the streamlines following a set of hyperboles. The velocity field from zero to infinity can be described by

$$U_x = ax \quad U_y = -ay \tag{15.47}$$

whereby the constant a could be arbitrary. If this asymptote is applied to a plate that is $2L$ wide then it is reasonable to assume that the streamlines in both directions will be symmetrical and that the fluid does not come from infinity but from a distance L. If the speed of the impacting fluid $-U_y = U$ is positioned at the distance L from the stagnation point then the estimate of the missing quantity should be

$$a \approx \frac{U}{L} \tag{15.48}$$

Because neither the static pressure nor the gravity apply, the distribution of the dynamic pressure will be obtained from the Bernoulli equation

$$p_0 + \frac{\rho}{2}U^2 = p + \frac{\rho}{2}\left(U_x^2 + U_y^2\right) \tag{15.49}$$

In the thin boundary layer it will be only

$$p = p_0 + \frac{\rho}{2}\left(U^2 - U_x^2\right) = p_0 + \frac{\rho U^2}{2}\left(1 - \frac{x^2}{L^2}\right) \tag{15.50}$$

so

$$-\frac{dp}{dx} = \frac{\rho U^2 x}{L^2} \tag{15.51}$$

Accordingly, the integral balance of momentum (15.40) remains,

$$\frac{d}{dx}\int_0^\delta u_x^2 dy - \frac{Ux}{L}\frac{d}{dx}\int_0^\delta u_x dy = \delta\frac{U^2 x}{L^2} - \frac{\mu}{\rho}\frac{\partial u_x}{\partial y}\bigg|_{y=0} \tag{15.52}$$

If for the boundary layer an assumption is made

$$u_x = \frac{Ux}{L}F(\eta) \quad \text{where} \quad \eta \equiv \frac{y}{\delta} \tag{15.53}$$

then a stepwise calculation shows that the thickness of the boundary layer does not depend on x. It comes out as

$$\delta = L\sqrt{\frac{\mu}{LU\rho}}\sqrt{\frac{\left.\frac{dF}{d\eta}\right|_{\eta=0}}{\int_0^1 \left(F-2F^2\right)d\eta + 1}} \tag{15.54}$$

The thickness of the Nernst type is

$$\delta_I = L\sqrt{\frac{\mu}{LU\rho}}\frac{1}{\sqrt{\left.\frac{dF}{d\eta}\right|_{\eta=0}\left[\int_0^1 \left(F-2F^2\right)d\eta + 1\right]}} \tag{15.55}$$

The accurate solution, after the substitution of the assumption (15.53) in the Navier-Stokes equation (Howarth, 1935), yields for the whole approach plane area a constant-thickness boundary layer

$$\delta_I = 0.811L\sqrt{\frac{\mu}{LU\rho}} = 0.811\sqrt{\frac{\mu}{xU_x\rho}} \tag{15.56}$$

and the friction factor is

$$f_x \equiv \frac{\mu\frac{U_x}{\delta_I}}{\frac{\rho}{2}U_x^2} = 2.466\sqrt{\frac{\mu}{xU_x\rho}} \tag{15.57}$$

As illustrated in Table 15.2 no major error will be committed by the use of various approximations of the profile function $F(\eta)$.

Table 15.2 Estimate of the boundary layer thickness for the flow close to the front stagnation point.

$F(y/\delta)$	$\delta_I/(LRe_L^{-1/2})$
Exact (Howarth)	0.811
y/δ	1.095
$2(y/\delta) - (y/\delta)^2$	0.913
$3/2(y/\delta) - 1/2(y/\delta)^3$	1.010
$4/3(y/\delta) - 1/3(y/\delta)^4$	1.045
$\sin(\pi/2y/\delta)$	1.000

Previously, the same outcome (except for the constant) was obtained for the axial impact of the fluid on a circular obstacle having a diameter $2L$ in cylindrical coordinates.

15.4.2 The flow past a rotating disc

Somewhat more complicated is the equation of flow in a half-space, the boundary of which is formed by a rotating disc (the **Karman problem**). When a characteristic length dimension is introduced by a relationship

$$r_c \equiv \sqrt{\frac{\mu}{\Omega \rho}} \tag{15.58}$$

then, at a position where $R \gg r_c$, the centrifugal force will apply and a boundary layer will get established in which with the growing distance from the plate, the peripheral velocity drops to almost zero. Simultaneously, the centrifugal force invokes a radial flow. Schematically, the nature of the flow around a rotating disc is depicted in Fig. 15.8.

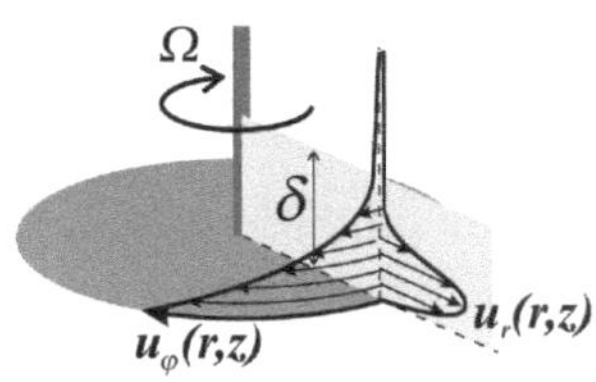

Fig. 15.8 A rotating disc at high Reynolds numbers.

The analysis of the flow shows that in this case the boundary layer thickness will be

$$\delta_I \approx 1.62 r_c \tag{15.59}$$

This exercise was elegantly solved by Theodor von Karman. The resulting velocity profiles are depicted in Fig. 15.9. This solution can be applied also to finite rotors so that it could be used for the depiction of flows around propellers, turbines, the rotors of pumps and mixers.

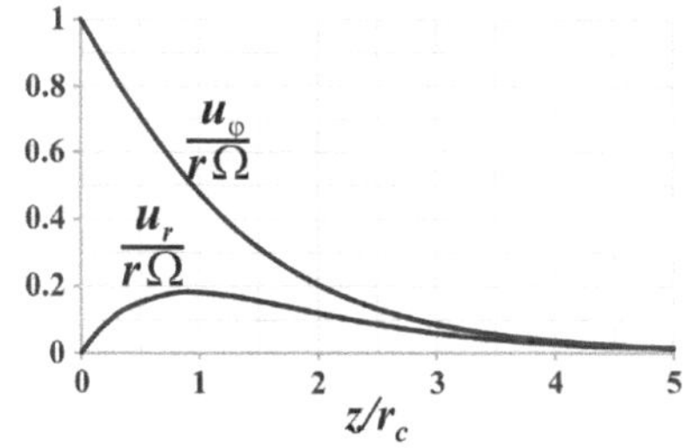

Fig. 15.9 Peripheral and angular velocities on a fast rotating disc.

15.5 The flow past realistic bodies

The flow past a body of finite thickness involves a combination of an inlet into a front plane with a constant thickness of the boundary layer (the Hiemenz problem) and a longitudinal flow past a body (the Prandtl problem) with the thickness of the boundary layer increasing. For slender bodies, Prandtl recommended introducing a local coordinate system with the coordinate x along the surface of the wall and the origin of the coordinates placed in the front critical point while the y coordinate being positioned at right angle to the wall. By combining this result with the distribution of the pressure along the wall calculated from the Euler equation, both problems become one. This fact became important for the study of aerodynamic problems like the design of the wing of an aircraft or of the blade of a propeller.

15.5.1 The flow past a sphere

The sphere represents one of the simplest spatial objects. In 1850, Sir George Stokes ventured to solve the flow past a sphere just after he derived the equation of motion including the influence of viscosity (currently known as the Navier-Stokes equation). The transformation of these equations into spherical coordinates makes it possible to define boundary conditions with regard to the observer who moves alongside with the sphere of a radius R at speed U in the infinite space.

$$\begin{aligned} u_s = 0 \ \text{ and } \ u_\theta = 0 \ \text{ for } \ s = R \\ u_s = U\cos\theta \ \text{ and } \ u_\theta = U\sin\theta \ \text{ for } \ s \to \infty \end{aligned} \tag{15.60}$$

15.5.1.1 The creeping flow

The solution of the Navier-Stokes equations without the terms on the left-hand side with the above boundary conditions leads to analytical expressions

$$\begin{aligned} u_s &= U\cos\theta\left[1 - \frac{3}{2}\left(\frac{R}{s}\right) + \frac{1}{2}\left(\frac{R}{s}\right)^3\right] \\ u_\theta &= U\sin\theta\left[1 - \frac{3}{4}\left(\frac{R}{s}\right) - \frac{1}{4}\left(\frac{R}{s}\right)^3\right] \end{aligned} \tag{15.61}$$

$$p = -\frac{3}{2}\frac{UR}{s^2}\cos\theta \tag{15.62}$$

as shown in Figs. 15.1 and 15.2.

Every element of the surface of the sphere is subject to a pressure p and a tangential stress

$$\tau_{s\theta} = \mu \frac{\partial u_\theta}{\partial s}\bigg|_{s=R} \tag{15.63}$$

For each of them it is possible to calculate the component that acts against the direction of the flow. Two resistance forces are obtained by their integration over the whole area of the sphere: one is derived from the tangential stress and is called the **friction resistance**

$$F_\tau = \int \tau_{R\theta} \sin\theta dS = 4\pi\mu UR \tag{15.64}$$

and the other is derived from the normal force (pressure) and is called the **form resistance**

$$F_n = \int p \cos\theta dS = 2\pi\mu UR \tag{15.65}$$

The expression for the **overall resistance force** is known as **the Stokes law**.

$$F = 6\pi\mu UR$$

Naturally, it is valid only at very low values of the Reynolds number

$$Re \equiv \frac{dU\rho}{\mu} \tag{15.66}$$

where $d = 2R$ is the diameter of the sphere. A drag coefficient C_D is usually introduced for the flow past a sphere so that the force is related to the front cross-sectional area and to the kinetic term of the Bernoulli equation. For a sphere having a diameter d

$$C_D \equiv \frac{F}{\frac{\pi d^2}{4}\frac{\rho U^2}{2}} = \frac{24}{Re} \tag{15.67}$$

Both the experimental experience and the complex mathematical analysis show that the assumption of a creeping flow (the idea about the negligibility of the inertial terms) is perfectly well acceptable for the Reynolds number $Re \ll 1$. At $Re = 0.2$ the inertial terms add about 5% of the overall resistance.

15.5.1.2 Stokes' paradox

It would appear that the side flow past an infinitely long cylinder in an unlimited space is a simpler task and that the solution to the equations if creeping flow is much easier to obtain than in the case of the sphere. It turns up, however, that there is not any steady solution in an unlimited space while leaving out all inertial terms. Oddly enough, it is sufficient to add a rough linearized estimate of the minuscule inertial forces (Ozeen model) and the solution becomes feasible.

15.5.1.3 The boundary layer for a flow past a sphere

The thickness of the boundary layer established on the surface of a sphere is controlled by relationships similar to the boundary layer on the surface of a flat plate. The magnitude of the local coefficient of friction is inversely proportional to the square root of the Reynolds number

$$f_x \sim Re_x^{-1/2} \tag{15.68}$$

The integration of its component in the direction of the bypassing flow yields a proportionality

$$C_D \sim Re_x^{-1/2} \tag{15.69}$$

By adding this relationship to the Stokes law (15.67) and by substituting the value 4 for the constant, the drag coefficient comes out as

$$C_D \approx \frac{24}{Re} + \frac{4}{Re^{\frac{1}{2}}} \tag{15.70}$$

The range of validity of this relationship for the flow past a sphere is not as wide as for the flow past a flat plate (it fits well only for $Re < 20$). The reason for this dichotomy lies in the possibility of the separation of the boundary layer.

15.5.1.4 Separation of the boundary layer, the wake

It is quite satisfactory to look at the distribution of pressure according to the Bernoulli equation for the potential inviscid flow past a sphere. Clearly, the lowest pressure must be in the vicinity of the fastest flow, that is, at the equator of the sphere. It then follows that a significant driving force appears on the rear side of the body acting against the direction of the flow. As a result, a **wake** is formed in which, at Reynolds numbers greater than one, a countercurrent circulation appears, initially as symmetric vortices. At higher

Reynolds numbers, the vortices get longer and the mismatch between the velocities of the ends of the vortices and the surrounding liquid generates instabilities. One vortex begins to extend faster and then separates. This process of vortex shedding is cyclically repeated. If a thin stream of a color tracer is injected into the stream (see Fig. 15.10) then a regular trail can be observed behind the bypassed body called the Karman vortex street.

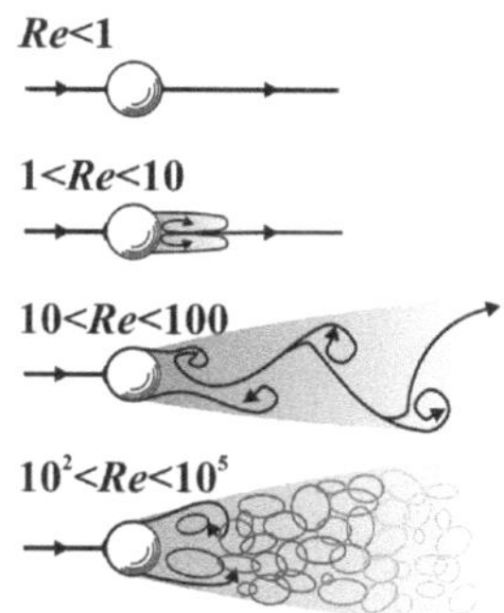

Fig. 15.10 Color trail behind a bypassed body and the formation of a wake.

Similar intermittent shedding of the boundary layer from a bypassed soft plane can be seen behind the flicker of a flag.

On further increasing the Reynolds number over 100, the large laminar vortices begin to split into smaller chaotic ones of the turbulent flow. This means that in this zone the condition of the potential irrotational flow is no longer met. The dynamic pressure on the incidence edge ($\Delta p \approx (\rho/2)U^2$) is not compensated by an opposite pressure on the rear side. Essentially, all the relevant energy is lost in the wake due to friction in the vortices. At higher Reynolds numbers, this energy exceeds all other viscous losses that obey the relationship (15.70). Therefore, most of the resistance to flow is due to the kinetic energy so the drag coefficient C_D remains essentially constant. This constant is characteristic for each geometrical setup. For a sphere it is, for example,

$$C_D \approx 0.44 \quad \text{for} \quad 500 < Re < 300{,}000 \tag{15.71}$$

For a wide range of variables this quantity can be added to relationship (15.70), so the resulting correlation is

$$C_D \approx \frac{24}{Re} + \frac{4}{Re^{\frac{1}{2}}} + 0.44 \tag{15.72}$$

This formula represents the experimental data sufficiently (Fig. 15.11).

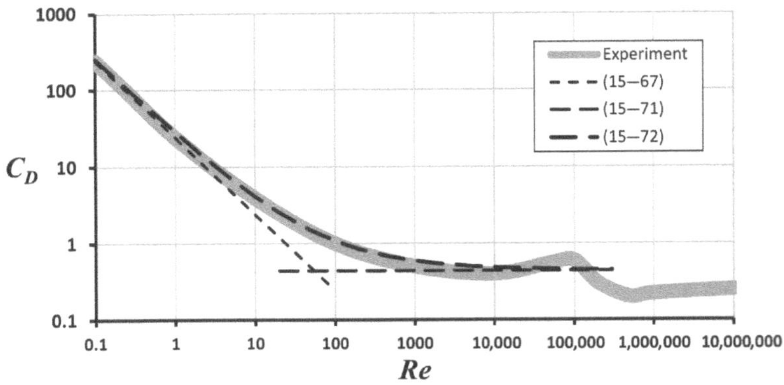

Fig. 15.11 Drag coefficient for a flow along a sphere: Experiments and the suggested correlations (15.67), (15.71), and (15.72).

The upper limit of the validity of Re is about 300,000 which corresponds to very rapid flows approaching locally supersonic speeds when wakes begin to peel off from the main body.

The values of the drag coefficients for bodies at high Reynolds numbers are presented in Table 15.3.

Table 15.3 Drag coefficient of some bodies at high Reynolds numbers.

The situation	C_D
The flow impinging at a right angle on a flat belt	2
A paratrooper falling with an open parachute	1.4
The flow across a cylindrical rod	1.2
The flow impinging at a right angle on a circular target	1.2
A bus	0.6
A sphere	0.44
An automobile	0.25
A drop like body	<0.1

It is remarkable that a body having the same frontal face but a smaller volume as a sphere (e.g., a circular target) can manifest a higher resistance to flow than the sphere itself. On the other hand, bodies having elongated shapes are not so much prone to the separation of the boundary layer and the loss of the mechanical energy as is the case with a sphere.

The pioneer of the aerodynamic research was N.J. Zhukovsky (1847–1921) who identified the optimal profile for an aircraft wing featuring rounded front plane and a sharp rear edge.

One of the methods of preventing the separation of the boundary layer is the suction of the boundary layer through a porous wall of the bypassed profile. This has been realized in some applications with a positive result.

Because the resistance to flow at high Reynolds numbers is proportional to the second power of speed so that the required mechanical input is proportional to the third power of speed, the resistance of air at low speeds may be negligible while at high speeds it could be the limiting factor.

15.6 The entry section of a pipe

Even though the result of the solution of the equation of motion of the laminar creeping flow in pipes satisfies the condition $Re < 2000$, a parabolic velocity profile could not be expected immediately after the entry of the fluid from a large container into a pipe. It is more likely to ignore the motion of the liquid in the container as well as any velocity distribution before the entry. Rather, fluid is more likely to enter the pipe as a plug and only after the entry it gets retarded by the friction on the walls.

Similarly, as is the case with a bypassed plate, a boundary layer will develop having an approximate thickness.

$$\delta \approx 5xRe_x^{-1/2} \tag{15.73}$$

Only when the boundary layers join in the center of the tube it may be expected that the parabolic profile was reached. From the requirement that $\delta \approx d/2$ an estimate of the length of the entry is obtained

$$x_{vstup} \approx 0.01\, d\, Re \tag{15.74}$$

This conclusion makes it possible to estimate whether the influence of the entry section on the flow could be neglected or not. The existence of an entry section increases somewhat the pressure loss in the flow. This is taken into account by the introduction of a correction which is essential when measuring the viscosity of liquids in short capillary viscometers.

If the liquid entering the pipe is already streamlined then the flow in the entry section of the pipe is almost laminar. Nevertheless, at higher Reynolds numbers the liquid will, in the end, resist the high shear rate at wall. Particularly, if some disorder (such as a local roughness) is initiated, the shearing will be replaced by rolling over and, further, changing to vortices that occupy increasingly more and more space behind the remaining thin

laminar sublayer, thus generating more and more vortices in opposite direction. The result is a developed **turbulent flow**.

15.6.1 The flow in narrow and wide channels

Problems similar to flow past bodies can be expected for flows in narrow and wide channels, in particular, the dependence on Reynolds number. At low Reynolds numbers, the flow is almost creeping; at higher Reynolds numbers a new boundary layer will be formed. When the channel widens further, the boundary layer will have a tendency to separate and form a wake in which most of the kinetic energy of the liquid will be lost.

One of the limiting cases is the flow into the almost limitless space of the fluid known as the free jet. A boundary layer is formed on the border of the jet that sets the remaining fluid in circular motion. The jet gets wider as it gets farther from the inlet and its velocity slowly even out with the surrounding fluid. It can be assumed that in finite space after widening the boundary layer will get separated and a substantial portion of the kinetic energy will, in the end, be lost in more or less regular vortices behind the widening.

15.7 Flow past bubbles and drops

15.7.1 The shape of fluid particles

The shape of bubbles and drops is due to the resultant of the acting forces. While the gravity tends to form horizontal boundaries, the interfacial tension has the tendency to minimize interfacial area by forming a spherical shape. The longitudinal scale that characterizes the influence of interfacial tension in the gravitational field is the Laplace length

$$d_p = \left(\frac{6V}{\pi}\right)^{\frac{1}{3}} \tag{15.75}$$

For the characteristic length of a particle the diameter of a sphere d_P that would have the same volume V as the particle is usually taken, thus

$$d_p = \left(\frac{6V}{\pi}\right)^{1/3} \tag{15.76}$$

The relative influence of forces acting on a fluid particle can be expressed by the Eotvos number

$$Eo \equiv \frac{d_p^2 |\Delta\rho| g}{\sigma} = \left(\frac{d_p}{L_L}\right)^2 \tag{15.77}$$

At $Eo < 1$ the interfaces are almost spherical, however, at $Eo > 10$ fluid particles are distinctly flattened; their maximum height is equal to $2L$ for asymptotic case of $Eo \rightarrow \infty$. The shapes of profiles of drops on an unwettable surface are shown in Fig. 15.12. The analogous case, but with an opposite orientation of the gravitational force, is shown by bubbles caught on a horizontal, perfectly wetted wall.

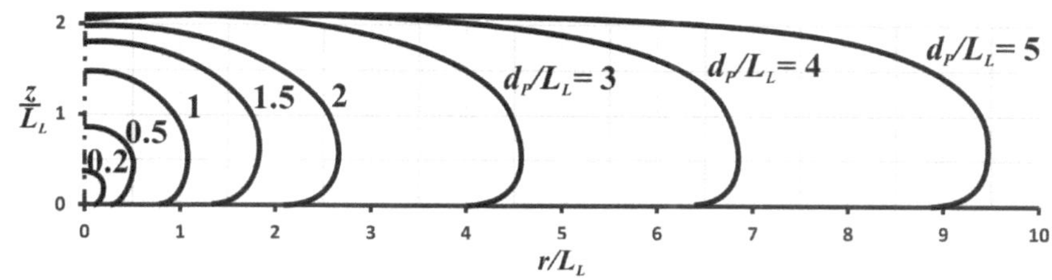

Fig. 15.12 Profiles of drops on an unwetted pad.

15.7.2 Rising bubbles, falling drops

The shape of an overflowed movable particle is affected as well. Small particles ($Eo < 1$) are almost spherical but the boundary conditions are different than those used for the Stokes solution (15.67). The equilibrium of tensile stresses on the interface of both phases generates circulation of the dispersed phase (index *d*) inside the particle. The motion of the particle relative to the surrounding fluid is easier. For bubbles in a liquid for which $\mu_d \ll \mu_c$, the friction coefficient of the creeping flow could be reduced down to

$$C_D = \frac{16}{Re} \tag{15.78}$$

In reality, however, the interfaces are blocked by surface active impurities so they behave as immobile. Larger rising or falling particles ($Eo > 1$) are influenced also by viscous and inertial forces expressed by the Reynolds number

$$Re_p \equiv \frac{d_p U \rho_c}{\mu_c} \tag{15.79}$$

where U is the relative velocity of the particle in relation to the surrounding continuous phase, the parameters of which carry the index c. A less common case is the creeping flow ($Re_p < 10$) when the viscous forces deform the particle giving it a vertically elongated shape.

A more common case is for $Re_p > 10$ when the viscous force of the medium does not play any significant role. The smaller fluid particle maintains its flattened shape as shown in Fig. 15.12 side regardless of whether it falls or rises.

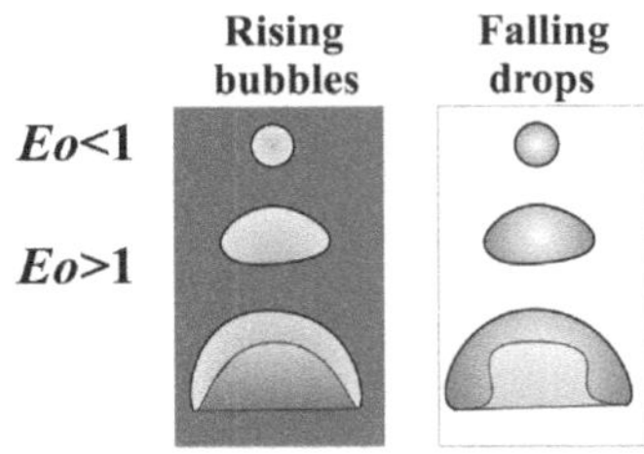

Fig. 15.13 Rising bubbles and falling drops.

While moving, the flattened particle is unstable leading to its rocking so the trajectory of its motion forms a wavy or a helical line. In still larger particles, a concave bulge forms (see Fig. 15.13). More pronounced oscillation of the shape and the motion results in the breakage of the particles into smaller objects.

It is worth noticing that the forces acting on a rain drop do not allow it to reach an aerodynamic shape. Such a shape will be attained only by a hanging drop in the moment of separation or by a tear sliding down on a partially wetted surface that shows signs of hysteresis (see Fig. 3.12).

15.8 Summary

For a two-dimensional flow, one of the Navier-Stokes equations is equal to zero. The remaining two equations of second order can be transposed by a mathematical manipulation to a single equation of the fourth order. This step might simplify the mathematical solution of the problem.

The complete solution of the velocity and pressure fields can be obtained for two border cases: viscous creeping flow and irrotational potential flow.

In real situations, the viscosity is often manifested only in limited regions, the so-called boundary layer, most often in the close vicinity of the wall. In this thin boundary layer, it is possible to simplify the Navier-Stokes equation by neglecting the less significant terms. This manipulation leads to the Prandtl equation of the boundary layer that is easier to solve. It further follows that, under these circumstances, the friction factor is inversely proportional to the second root of the Reynolds number. The calculated value depends only insignificantly on other assumptions made about the speed in the boundary layer.

Even though the concept of the boundary layer was developed for a flat plate simulating the wing of an aircraft, it is applicable for a large class of geometric configurations of flow.

The true existence of the boundary layer can be proved experimentally by studying the velocity field. However, when the flow bypasses the plate at very high Reynolds numbers, the boundary layer loses the character of a laminar flow and starts generating turbulent vortices.

When the flow bypasses realistic bodies, even at low Reynolds numbers, the boundary layer might separate and form first a laminar and then a turbulent wake. The friction in the turbulent wake dissipates most of the energy so the friction factor remains essentially constant.

The shape of bubbles and drops is due to the resultant of the acting forces. While the gravity tends to form horizontal boundaries, the interfacial tension has the tendency to minimize the interfacial area by forming a spherical shape. The dynamic pressure of the circumfluous fluid will flatten the liquid particle always on the downstream side regardless of whether it falls or rises. While moving, the flattened particle is unstable leading to its rocking so the trajectory of its motion forms a wavy or a helical line.

References

Howarth, L., 1935. ARC R&M 1632.
Prandtl, L., 1904. Reprinted in: NACA TM 452, 1928.

CHAPTER SIXTEEN

Extension of balances to turbulent flows

Contents

16.1 Termination of laminar flow

The turbulent flow will occur whether the fluid passes around a body or it flows inside tubes or channels. The critical Reynolds number is specific for each flow situation. The often claimed argument that it is just equal to 2100 is reasonably well valid only in the most ordinary straight tubes having a circular cross-section and smooth walls. It holds reasonably well in other similar flows but even in these situations, the state of turbulence can be achieved by inducing some additional disturbances, such as acoustic vibration. Inversely, sometimes it is possible to maintain the laminar flow in still flow situations in very smooth tubes up to the Reynolds number reaching 5000. This is obvious from the dependence of the friction coefficient Λ on the Reynolds number in Fig. 16.1.

The distribution of velocities in a pipe has been extensively measured and it turned out that, apart from the turbulent core in the central part of the tube, there always exists a laminar sublayer removed by a distance y from the wall in which the axial velocity is

$$u_z = y\frac{\tau_w}{\mu} \quad \text{for} \quad y < 5\frac{\mu}{\sqrt{\tau_w \rho}} \tag{16.1}$$

https://doi.org/10.1016/B978-0-12-818994-8.00016-6

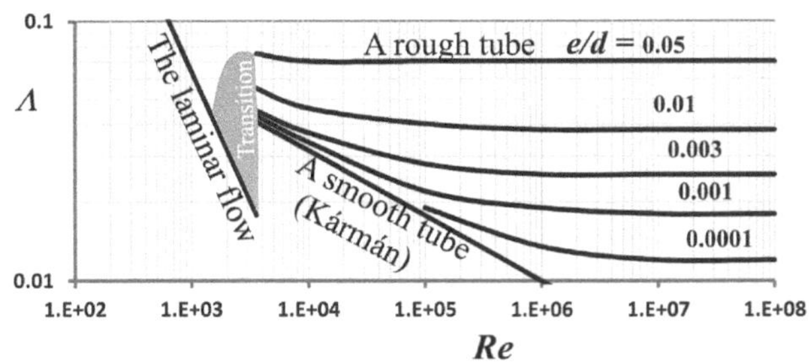

Fig. 16.1 The friction factor in a straight tube.
The theory and the experimental measurement of rough walls (*e* is the height of the roughness).

and which is linked to a preceding transient region in which the influence of viscosity is still pronounced

$$y < 30\frac{\mu}{\sqrt{\tau_w \rho}}$$

In the case of a unidirectional straight flow in a pipe, the inertial forces maintain the creeping character of flow up to the sudden collapse in the turbulent regime. In other situations, the inertial forces manifest themselves gradually, deep in the laminar regime as was shown, for example, on the instabilities of flow in Section 12.5. Then a gradual onset of turbulence can be expected even at lower Reynolds numbers.

It was already shown for the flow around a sphere (Fig. 15.10) that a regular wake begins to form already at $Re > 1$ and that this wake will show a turbulent nature at $Re > 100$. However, for the flow past a flat plate (Prandtl task), a distinct turbulent layer will not appear above the laminar boundary layer till $Re = 500{,}000$ (see Fig. 15.6).

Always, however, without regard to the geometrical setup, a narrow laminar and transient region remains at the boundary just as it happens in the pipe flow.

16.2 The solution of the equations of motion in the turbulent regime

Thus, each flow situation has a somewhat differently defined Reynolds number, a different role of the inertia, and a different nature of the potential disturbances in the laminar flow region. The value of the critical

Reynolds number can be found not only experimentally but also with the help of the theory of stability that requires, however, a challenging mathematical apparatus. Roughly speaking, some disturbance should be superimposed on a laminar flow and then this would be quickly suppressed if the Reynolds number were low. However, for more energetic flows above a certain bottom limit of the Reynolds number, it turns out that the disturbances of a certain kind and size get magnified. Above the upper limit of the critical Reynolds number, the solutions of the equations of motion are characterized by bifurcations. Any disturbance introduced in the realistic flows will be enhanced and the flow under all circumstances reaches a chaotic turbulent nature. This is, however, a deterministic chaos (governed by equations) that follows certain laws of statistics.

In the turbulent regime of flow, eddies get created, and they last for a certain time and then they cease to exist. They have diverse sizes and they last for different lengths of time. It is not surprising that this problem was the subject of interest to A.N. Kolmogorov, who is the founder of the modern theory of probability. According to this theory, a whole set of vortices gets created. The Reynolds number of the largest ones is so high that the influence of the inertia is predominant. Only the small vortices are responsible for the viscous dissipation of energy. The boundary conditions can be derived from the **intensity of dissipation of the turbulent energy** ε related to the unit of mass [J/kg $\equiv$ m^2/s^2] with the use of the kinematic viscosity. Thus, it turns out that the **Kolmogorov's scale of turbulence** may be

$$\textbf{Longitudinal scale}\quad \eta \equiv \frac{\nu^{3/4}}{\varepsilon^{1/4}} \tag{16.2}$$

$$\textbf{Temporal scale}\quad t_\eta \equiv \left(\frac{\nu}{\varepsilon}\right)^{1/2} \tag{16.3}$$

$$\textbf{Speed base}\quad u_\eta \equiv (\nu\,\varepsilon)^{1/4} \tag{16.4}$$

The local intensity of dissipation can be roughly estimated from the measured integral dynamic quantities (such as the pressure loss or the mechanical power input).

The currently available computational software allows for the solution of the Navier-Stokes equations and also for unsteady-state processes. In principle, the problems of turbulence can be studied numerically by the method **DNS** (**d**irect **n**umerical **s**imulation), since it also automatically includes the origin and the extinction of the vortices when the procedure, while proceeding through the bifurcations, accidentally picks up any one of the routes. To obtain a realistic solution, the characteristic dimension of the chosen elements of space and time must be smaller than the Kolmogorov's scale. On the other hand, the examined time element must be long enough to allow for the statistical processing of the results. Apparently, the volumes of the calculations are so large that, so far, only some simple demonstration calculations have been performed.

16.3 The averaged flow. Reynolds stresses

In the following, an analysis of the velocity field for a turbulent flow will be performed that makes it possible to summarize some properties of the spatial whirling into scalar quantities, estimate their dependence on the geometry of the flow situation, and derive relationships governing the specific situation.

The oscillation of the velocity caused by the formation of eddies is a typical phenomenon of turbulent flow. The information about the mean velocity can be obtained by taking an average of the velocities over a sufficiently long time. From there, it is then possible to determine the flow rate of the fluid through a chosen area. The instantaneous velocity component u in the direction of flow i can be split into two parts: the **averaged value** $\overline{u_i}$ and the **oscillating component** u_i

$$u_i = \overline{u_i} + u_i' \tag{16.5}$$

The averaging of the quantities u_i' and of their derivatives $\partial u_i'/\partial j$, $\partial^2 u_i'/\partial j^2$, $\partial u_i'/\partial t$ will always result in zero. If the Navier-Stokes equation is expressed in terms of quantities $\overline{u_i} + u_i'$and averaged, then a relationship for the component of the momentum balance in the x direction is obtained in Cartesian coordinates in which three more terms appear, which correspond to the averaged product of arbitrary quantities

$$\rho\left(\frac{\partial \overline{u_x}}{\partial t} + \overline{u_x}\frac{\partial \overline{u_x}}{\partial x} + \overline{u_x'\frac{\partial u_x'}{\partial x}} + \overline{u_y}\frac{\partial \overline{u_x}}{\partial y} + \overline{u_y'\frac{\partial u_x'}{\partial y}} + \overline{u_z}\frac{\partial \overline{u_x}}{\partial z} + \overline{u_z'\frac{\partial u_x'}{\partial z}}\right)$$
$$= \rho g_x - \frac{\partial \overline{p}}{\partial x} + \mu\left(\frac{\partial^2 \overline{u_x}}{\partial x^2} + \frac{\partial^2 \overline{u_x}}{\partial y^2} + \frac{\partial^2 \overline{u_x}}{\partial z^2}\right) \tag{16.6}$$

This relationship is usually modified by moving the additional members to the right-hand side of the equation and by assigning it to an acronym **RANS** (**R**eynolds **a**veraged **N**avier-**S**tokes)

$$\rho\left(\frac{\partial\overline{u_x}}{\partial t}+\overline{u_x}\frac{\partial\overline{u_x}}{\partial x}+\overline{u_y}\frac{\partial\overline{u_x}}{\partial y}+\overline{u_z}\frac{\partial\overline{u_x}}{\partial z}\right)$$
$$=\rho g_x-\frac{\partial\overline{p}}{\partial x}+\frac{\partial}{\partial x}\left(\mu\frac{\partial\overline{u_x}}{\partial x}+\rho\overline{u_x{}'u_x{}'}\right)+\frac{\partial}{\partial y}\left(\mu\frac{\partial\overline{u_x}}{\partial y}+\rho\overline{u_x{}'u_y{}'}\right)$$
$$+\frac{\partial}{\partial z}\left(\mu\frac{\partial\overline{u_x}}{\partial z}+\rho\overline{u_x{}'u_z{}'}\right) \tag{16.7}$$

Because the expressions $\mu\left(\frac{\partial\overline{u_i}}{\partial j}\right)$ represent in reality the averaged components of the tensile stress, with them are equally labeled groups added in parallel

$$\tau_T\equiv\rho\,\overline{u_i{}'u_j{}'} \tag{16.8}$$

as so-called **Reynolds stresses** contributing to the dissipation of the mechanical energy.

16.3.1 Scalar parameters of a turbulent flow

In order to get the Reynolds stresses formally the same dependence as they have the stresses of the laminar flow, an **eddy viscosity** μ_T was proposed as follows:

$$\rho\,\overline{u_i{}'u_j{}'}=\mu_T\left(\frac{\partial\overline{u_i}}{\partial j}+\frac{\partial\overline{u_j}}{\partial i}\right) \tag{16.9}$$

It is then assumed that it depends on several scalar quantities. They are as follows:

the intensity of the dissipation of the turbulent energy ε [W/kg], calculated as

$$\varepsilon\equiv\frac{\mu}{2\rho}\sum\overline{\left(\frac{\partial u_i{}'}{\partial j}+\frac{\partial u_j{}'}{\partial i}\right)^2} \tag{16.10}$$

the turbulent kinetic energy k, that is the energy of the vortices,

$$k\equiv\frac{1}{2}\sum\overline{u_i'^2} \tag{16.11}$$

the specific rate of dissipation ω [s^{-1}], where

$$\omega\equiv\frac{\rho k}{\mu_T} \tag{16.12}$$

It can be expected that for the turbulent viscosity, the following proportionality will apply

$$\mu_T = C_\mu \frac{k^2}{\varepsilon} \tag{16.13}$$

where the constant of proportionality C_μ is equal to about 0.09.

Implicit equations were proposed for the determination of the quantities k, ε, and ω that are used by computational software. They contain a number of empirical constants probably derived from experimental measurements. The values of the averaged and of the oscillating velocity components can be determined in specific situations by the statistical analysis of the data on instant local velocities obtained by measurement in transparent liquids using methods HWA, LDA, and PIV (see Section 5.1). The results of these measurements are incorporated into the software packages. Usually, to solve velocity fields, a combined strategy is used whereby the ***k*-*ω* model** is used close to the wall while the ***k*-*ε* model** is used for the free flow. Nevertheless, for applications to unusual hydrodynamic situations, it is advisable to compare at least some of the partial results with the experiments to increase the credibility of the solution.

16.4 The influence of large eddies

The technique of **LES** (**L**arge **E**ddy **S**imulation) is other option than the RANS method. In this case, the momentum transfer is solved for a velocity distribution stripped of the smallest eddies. Principally, this approach is close to the Brandt model the key point of which was the definition of the mixing length L_T corresponding to the characteristic size of those eddies that control the momentum transfer. This notion stems from the image that the eddies transport the momentum in a similar fashion as it happens in the random motion of molecules in gases. It was assumed that as the kinematic viscosity is, according to Eq. (2.9), proportional to the product of the mean free path of the molecules and their velocity it was assumed that there is an isotropic turbulent viscosity μ_T that depends on the product of the mixing length and the characteristic oscillating velocity component for which it can be expected that

$$u' \sim L_T \left|\nabla \bar{\boldsymbol{u}}\right| \tag{16.14}$$

$$\text{so that } \mu_T/\rho \sim L_T^2 \left|\nabla \bar{\boldsymbol{u}}\right| \tag{16.15}$$

It turned out, however, that the mixing length L_T is not a function of the material parameters and of the Reynolds number only. It becomes smaller in the vicinity of the walls. If this model is used for the numerical simulation of the velocity fields then the elements in the boundary layer close to the wall must be chosen adequately small. Again, a large number of commercial strategies are available for numerical solutions. In principle, it can be expected that the LES simulation techniques, when correctly applied, offer more accurate results than the RANS methods, however, for the price of higher demands on the computer time. And yet, they are substantially more economical than the DNS techniques.

16.5 The turbulent transfer of heat and mass

In the same way as the random components of the particle velocity influence the transfer of momentum, they also affect the heat and mass transfer. Similarly, the **eddy thermal conductivity** is introduced with the help of the Prandtl number

$$k_T = \frac{\mu_T \, c_P}{\mathrm{Pr}} \tag{16.16}$$

and further the **eddy diffusivity** with the help of the Schmidt number

$$\mathcal{D}_T = \frac{\mu_T}{\rho \, Sc} \tag{16.17}$$

Also, these quantities may in some cases facilitate the interpretation of experimental data. Unfortunately, a prediction that would allow the construction of reliable mathematical models of processes in the turbulent regime is missing.

16.6 Summary

A distinct unsteady whirling will appear in flowing liquids at high Reynolds numbers under steady conditions that are known as the turbulence. The critical Reynolds number depends on the geometrical setup. It could be several orders of magnitude higher or lower than the value of about 2000 for a straight pipe. Turbulence is not at variance with the equations of motion. These equations can be mathematically characterized as being nonlinear. If the nonlinearity is sufficiently high, they do not lead to unambiguous solutions. This is manifested as deterministic chaos.

Local velocities and their oscillations were already experimentally determined for several geometrical set-ups. Detailed velocity and stress distributions in turbulent flows can also be obtained by the numerical solution of the Navier-Stokes equations of unsteady flow. It does, however, require a large volume of computation.

Somewhat more feasible is the simulation that neglect in each element the smaller of the local eddies and respects only the larger.

Substantially simpler is used to study the averaged steady flow where the eddies are manifested by additional Reynolds stresses. Those depend on scalar invariants of the turbulent flows that can be obtained by the solution of the implicit balance equations complemented by several empirical terms. Recommended software procedures have been developed on this basis.

Large eddies that transport the momentum in the turbulent flow can be characterized by a mixing length and by the turbulent viscosity derived from it. These eddies transport heat and mass by the same mechanism.

CHAPTER SEVENTEEN

Solution of the equation of convection

Contents

17.1 Heat transfer in a pipe

The basic problem to be solved for a pipe flow is the case when the fluid that has a constant temperature and a fully developed velocity profile enters a section where the wall temperature suddenly and sharply changes (jump change). Cylindrical coordinates are used to describe the flow in a tube of a circular cross section, the origin of which is positioned in the spot of the jump change of the wall temperature. The radius of the tube $R = d/2$ is used as the characteristic longitudinal dimension in place of the diameter d. Thus, the velocity profile $u_z(r)$ is determined by these specifications.

Transport and Surface Phenomena
https://doi.org/10.1016/B978-0-12-818994-8.00017-8

In a steady state the first derivative by time $\partial/\partial t=0$.

The **Fourier-Kirchhoff equation** will then assume the form

$$u_z(r)\frac{\partial T}{\partial z}=a\left(\frac{\partial^2 T}{\partial r^2}+\frac{1}{r}\frac{\partial T}{\partial r}+\frac{\partial^2 T}{\partial z^2}\right) \tag{17.1}$$

Next, an analysis of the significance of the individual terms of this equation is performed:

$$\begin{aligned} u_z(r)\frac{\partial T}{\partial z}&=o\left(U\frac{\Delta T}{z}\right)\\ k\frac{\partial^2 T}{\partial z^2}&=o\left(k\frac{\Delta T}{z^2}\right) \end{aligned} \tag{17.2}$$

where z is the characteristic length of the section where the heat is exchanged. The ratio of the magnitudes of the individual terms is

$$\frac{a\dfrac{\partial^2 T}{\partial z^2}}{u_z(r)\dfrac{\partial T}{\partial z}}=o\left(\frac{a}{Uz}\right) \tag{17.3}$$

The expression in the brackets on the right-hand side of this equation can also be written as

$$\frac{a}{Uz}=\frac{d}{z}\frac{a}{Ud}\equiv\frac{d}{z}\frac{1}{RePr} \tag{17.4}$$

in which a new dimensionless number appears, the product of the Reynolds number and of the Prandtl number, known as the **Peclet number**

$$Pe\equiv RePr \tag{17.5}$$

In most engineering tasks on heat transfer, this number is larger by several orders of magnitude than one, so that the term with the second derivative by z can be neglected (with the exception of an insignificant region of very small z in the vicinity of the temperature jump). This means that, in the longitudinal direction, the convection of heat is much more significant than the conduction. Therefore, the equation to be solved is

$$u_z(r)\frac{\partial T}{\partial z}=a\left(\frac{\partial^2 T}{\partial r^2}+\frac{1}{r}\frac{\partial T}{\partial r}\right) \tag{17.6}$$

17.2 The solution for a constant wall temperature

The solution depends on the boundary conditions. One of the typical cases is a presumption that the temperature T_0 is constant at the entrance of the fluid in the section where the heat transfer occurs, thus

$$\text{for } z = 0 \text{ and } 0 < r < R \quad \text{the } T = T_0 \tag{17.7}$$

In this section, the wall temperature T_W is different, thus

$$\text{for } z > 0 \text{ and } r = R \quad \text{the } T = T_W \tag{17.8}$$

Further, the condition of symmetry applies

$$\text{for } r = 0 \text{ is } \partial/\partial r = 0 \tag{17.9}$$

Formally, the solution is similar to the case of the heat conduction in a finite space. If this equation is transformed in a dimensionless form by the introduction of new variables

$$\eta \equiv \frac{R - r}{R} \tag{17.10}$$

$$\zeta \equiv \frac{za}{UR^2} = 4\left(\frac{z}{d}\right)\frac{1}{RePr} \tag{17.11}$$

$$\theta_1(\eta, \zeta) \equiv \frac{T(r \cdot z) - T_0}{T_W - T_0} \tag{17.12}$$

$$w(\eta) \equiv \frac{u_z(r)}{U} \tag{17.13}$$

a dimensionless equation will be obtained

$$w(\eta)\frac{\partial \theta_1}{\partial \zeta} = \frac{\partial^2 \theta_1}{\partial \eta^2} + \frac{1}{1-\eta}\frac{\partial \theta_1}{\partial \eta} \tag{17.14}$$

with the following boundary conditions:

$$\text{for } \zeta = 0 \text{ and } 0 < \eta < 1 \quad \text{is } \Theta_1 = 0 \tag{17.15}$$

$$\text{for } \zeta > 0 \text{ and } \eta = 0 \quad \text{is } \Theta_1 = 1 \tag{17.16}$$

and with the condition of symmetry being

$$\text{for } \eta = 1 \quad \text{is } \partial/\partial \eta = 0 \tag{17.17}$$

Specifically, for the laminar flow of a Newtonian fluid in a tube of a circular cross section it will be

$$w(\eta) = 4\eta - 2\eta^2 \tag{17.18}$$

This partial differential equation belongs to the class of **Sturm-Liouville equations** the solution of which can be reached by the method of the separation of variables (see Section 21.2). The result can be expressed by an infinite series

$$\theta_1 = 1 - \sum_{n=1}^{\infty} A_n F_n(\eta) \exp\left(-b_n^2 \zeta\right) \tag{17.19}$$

For a given velocity distribution given by Eq. (17.18), the eigenfunctions $F_i(\eta)$ cannot be expressed by the elementary functions or by the usual tabulated special functions. Therefore it is necessary to find them numerically along with the eigenvalues b_n^2. The appropriate eigenfunctions are shown in Fig. 17.1.

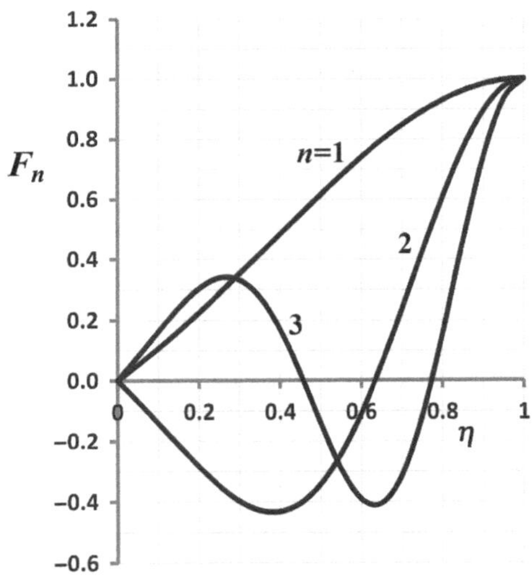

Fig. 17.1 First three eigenfunctions for the flow in a circular tube having a constant wall temperature.

Qualitatively, they are somewhat reminiscent of the sinus function in the interval of $\langle 0;\ (2n-1)\pi/2 \rangle$.

17.2.1 The solution for high values of ζ

Because the succession of the eigenvalues increases

$$b_1^2 = 3.656793\ldots,\quad b_2^2 = 22.305\ldots,\quad b_3^2 = 55.960\ldots \tag{17.20}$$

an important result is the asymptotic relationship

$$\lim_{\zeta\to\infty}\theta_1 = 1 - A_1 F_1(\eta)\exp\left(-b_1^2\zeta\right) \tag{17.21}$$

showing how the solution is converging toward the final value of $\Theta_1 = 1$. Here, the multiplier constant is

$$A_1 = 1.476435\ldots \tag{17.22}$$

17.2.2 The solution for low values of ζ

Similarly in the case of unsteady heat conduction for small values of $\zeta \to 0$, that is, in those places where the fluid remains only for a short while, the overall heat conduction occurs only in a thin layer close to the wall. The velocity profile in this thin layer can be approximated by the first term of the Taylor series

$$\lim_{r\to R} u_z(r) = \frac{U(R-r)}{R} w' \tag{17.23}$$

where w' is the dimensionless shear rate at the wall

$$w' \equiv \frac{R}{U}\left(-\frac{du_z}{dr}\right)\bigg|_{r=R} \tag{17.24}$$

This can be expressed in dimensionless terms as

$$\lim_{\zeta\to 0} w(\eta) = \eta w' \tag{17.25}$$

where just in laminar flow $w' = 4$.

Further, it will suffice to keep in mind that the wall temperature in the region of heat conduction drops from $\Theta_1 = 1$ to $\Theta_1 = 0$. It is, therefore, possible to view the solution as if it were valid from 0 to infinity where

$$\lim_{\eta\to\infty}\theta_1 = 0 \tag{17.26}$$

Simultaneously, it is possible to ignore the curvature of the tube where the heat transfer takes place and the task is reduced to a simple solution of an equation

$$\eta w' \frac{\partial\theta_1}{\partial\zeta} = \frac{\partial^2\theta_1}{\partial\eta^2} \tag{17.27}$$

Lévêque suggested a way to solve this equation by the use of a dimensionless combined variable

$$\xi \equiv \eta \left(\frac{w'}{9\zeta} \right)^{1/3} \tag{17.28}$$

that leads to a situation mathematically analogous to the solution of heat conduction to a half-space (Section 14.2).

17.2.2.1 The derivation

The substitution of Eq. (17.28) in the equation of heat convection (17.27) results, after an adjustment, in an ordinary differential equation

$$\frac{d^2\theta_1}{d\xi^2} + 3\xi^2 \frac{d\theta_1}{d\xi} = 0 \tag{17.29}$$

With the use of the appropriate boundary conditions

$$\theta_1 = 1 \quad \text{for}\, \xi = 0 \tag{17.30}$$

$$\theta_1 = 0 \quad \text{for}\, \xi \rightarrow \infty \tag{17.31}$$

the solution is obtained in the form

$$\theta_1(\xi) = 1 - \frac{1}{\Gamma\left(\frac{4}{3}\right)} \int_0^{\xi} \exp\left(-p^3\right) dp \tag{17.32}$$

where p is an auxiliary variable. The relevant value of the ordinary gamma function is

$$\Gamma\left(\frac{4}{3}\right) \equiv \int_0^{\infty} \exp\left(-p^3\right) dp = 0.892979\ldots \tag{17.33}$$

17.2.2.2 The result

The Leveque solution, when transposed back in the dimensional variables, can be expressed as

$$T = T_0 + (T_w - T_0)\, \Psi_1\left(\frac{R - r}{\Delta_1} \right) \tag{17.34}$$

where the boundary layer thickness is

$$\Delta_1 = 1.8575 \left(\frac{z\, a}{w'} \right)^{1/3} \tag{17.35}$$

where the respective function $\Psi_1((R-r)/\Delta_1)$ is graphically shown in Fig. 17.2.

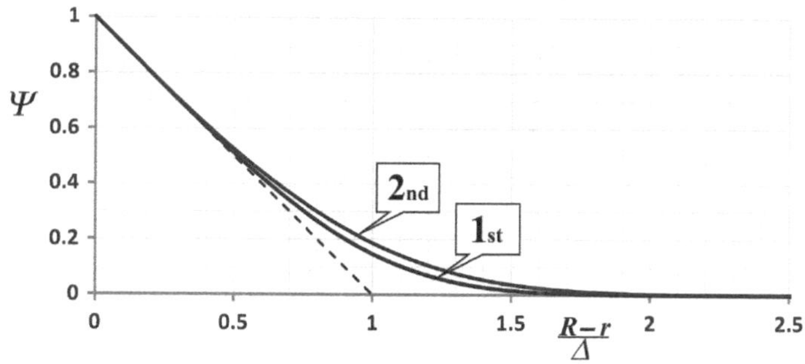

Fig. 17.2 The dimensionless temperature profile at low values ζ for the first boundary conditions (17.8) and for the second case (17.62).

At this, the intensity of the local heat flow through the wall is

$$q(z) \equiv k \left.\frac{\partial T(r,z)}{\partial r}\right|_{r=R} = \frac{k(T_0 - T_W)}{\Delta_1} \tag{17.36}$$

and

$$\Delta(z) = 1.8575 \left(\frac{w'}{z\,a}\right)^{1/3} \tag{17.37}$$

17.2.2.3 The generalization

Without going into the details, it is possible to find that in the case of a variable $w' = w'(z)$, the result, under otherwise equal conditions, is

$$q(z) = 0.53837 \frac{k(T_0 - T_W)}{a^{1/3}} \frac{1}{z} \left(\int_0^z \sqrt{w'(z)} dz \right)^{2/3} \tag{17.38}$$

17.3 The Nusselt number

The mean temperature of flow varies with the growing distance along the length of the pipe. As a rule, a mean mixed temperature T_m is defined as being the temperature that would result from the blending of the stream progressing through a given surface. In a sense, it is an expansion of the definition (1928) that was used for a stationary medium. Accordingly, it will be

$$T_M(z) \equiv \frac{\displaystyle\int_0^S \rho c_P u_z T dS}{\displaystyle\int_0^S \rho c_P u_z dS} \tag{17.39}$$

Specifically, for a round tube

$$T_M(z) \equiv \frac{\displaystyle\int_0^R \rho c_P u_z(r) T(r, z) 2\pi r dr}{\displaystyle\int_0^R \rho c_P u_z(r) 2\pi r dr} \tag{17.40}$$

that in the case of constant material properties becomes

$$T_M(z) \equiv \frac{\displaystyle\int_0^R r u_z(r) T(r, z) dr}{\displaystyle\int_0^R r u_z(r) dr} \tag{17.41}$$

The change of T_M along the length of the tube is related to the intensity of the heat flow q through the wall in a given location

$$\pi R^2 U \rho c_P \frac{dT_M}{dz} = 2\pi R q(z) = 2\pi R k \left.\frac{\partial T(r, z)}{\partial r}\right|_{r=R} \tag{17.42}$$

The intensity of the heat flow q through the wall in a given location is

$$q(z) = k \left.\frac{\partial T(r, z)}{\partial r}\right|_{r=R} \tag{17.43}$$

so that it will also be valid

$$q(z) = \frac{\pi R^2 U \rho c_P}{2\pi R} \frac{dT_M}{dz} \tag{17.44}$$

The coefficient of heat transfer varies along the distance z. It is, therefore, useful to define a **local coefficient of heat transfer** as

$$h_z(z) \equiv \frac{q(z)}{\Delta T(z)} \tag{17.45}$$

whereby for the characteristic temperature difference will be chosen

$$\Delta T(z) = T_W(z) - T_M(z) \tag{17.46}$$

Then the **local Nusselt number** will be

$$Nu_z(z) \equiv \frac{h_z(z)d}{k} = \frac{2R\left.\dfrac{\partial T(r,z)}{\partial r}\right|_{r=R}}{T_W(z) - T_M(z)} = \frac{\dfrac{R^2 U\rho c_P}{\lambda}\dfrac{dT_M}{dz}}{T_W(z) - T_M(z)} \tag{17.47}$$

which transposes in dimensionless numbers as

$$Nu_z(\zeta) = \frac{2\left.\dfrac{d\theta_1(\eta,\zeta)}{d\eta}\right|_{\eta=0}}{\theta_{1W}(\zeta) - \theta_{1M}(\zeta)} = \frac{\dfrac{d\theta_{1M}}{d\zeta}}{\theta_{1W}(\zeta) - \theta_{1M}(\zeta)} \tag{17.48}$$

It is often useful to know how much of the heat is transferred in the whole section between the temperature jump and the final distance z. Therefore a **mean coefficient of heat transfer** is defined as

$$\overline{h}(z) \equiv \frac{\displaystyle\int_0^z q(z)dz}{2\pi Rz\overline{\Delta T|_0^z}} = \frac{\pi R^2 U\rho c_P[T_M(z) - T_M(0)]}{2\pi Rz\overline{\Delta T|_0^z}} \tag{17.49}$$

whereby for the characteristic temperature difference the usually chosen arithmetic means

$$\overline{\Delta T|_0^z} \equiv \frac{[T_W(0) - T_M(0)] + [T_W(z) - T_M(z)]}{2} \tag{17.50}$$

The mean Nusselt number then turns out to be

$$\overline{Nu}(z) = 2R\frac{\pi R^2 U[T_M(z) - T_M(0)]}{2\pi aRz\overline{\Delta T|_0^z}} \tag{17.51}$$

Generally, the Nusselt number so defined can be calculated from experimentally determined quantities; if its magnitude is known it is then possible to determine the state of the fluid leaving the specific section of the heat exchanger. In specific cases, when simple boundary conditions apply then it is possible to calculate the Nusselt number and the mean temperature on theoretical grounds.

17.3.1 A round tube having a constant wall temperature

In this particular case when applying the boundary conditions (17.15) and (17.16), then the equations in the dimensionless format will be

$$\theta_M(\zeta) \equiv \frac{T_M(z) - T_0}{T_W - T_0} = \int_0^1 (1-\eta)w(\eta)\theta_1(\eta,\zeta)d\eta \tag{17.52}$$

$$Nu_z(\zeta) \equiv \frac{2\left.\dfrac{d\theta(\eta,\zeta)}{d\eta}\right|_{\eta=0}}{\theta_{1W}(\zeta) - \theta_{1M}(\zeta)} = \frac{\dfrac{d\theta_{1M}}{d\zeta}}{\theta_{1W}(\zeta) - \theta_{1M}(\zeta)} = \frac{\dfrac{d\theta_{1M}}{d\zeta}}{1 - \theta_{1M}(\zeta)} \tag{17.53}$$

and

$$\begin{aligned}\overline{Nu}(\zeta) &\equiv \frac{2[\theta_{1M}(\zeta) - \theta_{1M}(0)]}{\zeta[\theta_{1W}(0) - \theta_{1M}(0) + \theta_{1W}(\zeta) - \theta_{1M}(\zeta)]} \\ &= \frac{2\theta_{1M}(\zeta)}{\zeta[2 - \theta_{1M}(\zeta)]}\end{aligned} \tag{17.54}$$

The asymptotes for a short heat exchange area (low values of ζ):

$$\lim_{\zeta\to 0} Nu_z(\zeta) = \frac{2}{0.893}\left(\frac{w'}{9\zeta}\right)^{1/3} = 1.077\left(\frac{x}{d}\right)^{-1/3} Re^{1/3} Pr^{1/3} \tag{17.55}$$

$$\lim_{\zeta\to 0} \overline{Nu}(\zeta) = \frac{3}{2} Nu_z(\zeta) = 1.615\left(\frac{x}{d}\right)^{-1/3} Re^{1/3} Pr^{1/3} \tag{17.56}$$

(These values are somewhat overstated in view of the neglect of the curvature of the pipe wall.)

The asymptotes for a long heat exchange area (high values of ζ):

$$\lim_{\zeta\to\infty} \theta_M(\zeta) = 1 - A_1 \int_0^1 2\eta F_1(\eta) d\eta \exp\left(-b_1^2\zeta\right) \tag{17.57}$$

$$\lim_{\zeta\to\infty} Nu_z(\zeta) = b_1^2 = 3.567 \tag{17.58}$$

It is of less interest that for a given choice of the mean driving force (17.51) it is

$$\lim_{\zeta\to\infty} \overline{Nu}(\zeta) = \frac{2}{\zeta} = \frac{1}{2}\frac{d}{z} Re Pr \tag{17.59}$$

which means that the temperature of the stream changes only slightly while the area for the heat transfer further grows. The exact solution and the asymptote of the local and of the mean Nusselt number are shown in Fig. 17.3. It is obvious from there that the two asymptotes suffice by themselves to provide quite a tight description of the process.

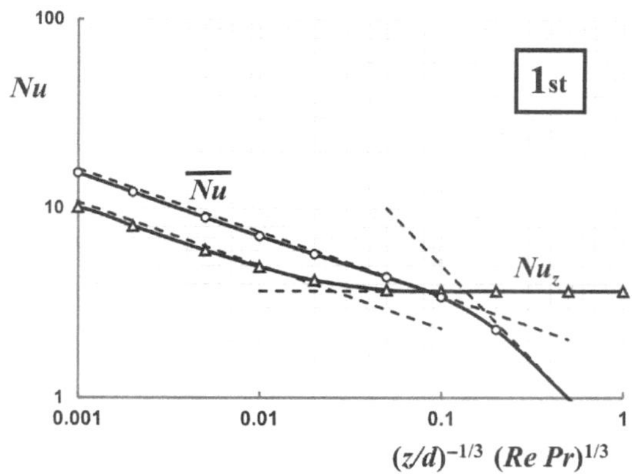

Fig. 17.3 The local and the mean Nusselt number for the flow in a circular tube having a constant wall temperature; the exact solution and the asymptotes (17.54), (17.57), (17.55), and (17.59).

17.4 The solution for constant heat flow through a wall

Here too an assumption is made that in the entrance to the heat exchange section the temperature T_0 is constant so that

$$\text{for } z = 0 \text{ and } 0 < r < R \quad \text{is } T = T_0 \tag{17.60}$$

and the symmetry condition

$$\text{for } r = 0 \quad \text{is } \partial/\partial r = 0 \tag{17.61}$$

It is further assumed that the intensity of the heat flow q_W through the wall is constant. Thus

$$\text{for } z > 0 \text{ and } r = R \quad \text{it will be } q = q_W, \text{ i.e., } \partial T/\partial r|_{r=R} = q_W/k \tag{17.62}$$

Formally, the solution is very similar to that of the heat conduction in a finite space (Section 14.4).

If this equation is transposed in a dimensionless form by using a variable

$$\theta_2 \equiv \frac{(T - T_0)k}{2Rq_W} \tag{17.63}$$

then

$$\lim_{\zeta \to \infty} \theta_2(\zeta) \equiv \zeta + \frac{11}{48} - \frac{1}{2}\eta - \frac{1}{4}\eta^2 + \frac{1}{2}\eta^3 - \frac{1}{8}\eta^4 \tag{17.64}$$

and then

$$\lim_{\zeta\to\infty} Nu_z(\zeta) = \frac{48}{11} = 4.36\overline{36} \tag{17.65}$$

A complete solution of this problem can be expressed in the form of an infinite series that converges fast for higher values of ζ.

$$\theta_2(\eta,\zeta) = \lim_{\zeta\to\infty}\theta_2(\zeta) - \sum_{n=1}^{\infty} A_n F_n(\eta)\exp\left(-b_n^2\zeta\right) \tag{17.66}$$

In this particular case, the functions $F_n(\eta)$ is somewhat reminiscent of the cosine function in the interval $\langle 0;\, n\pi\rangle$.

For small values of ζ, the solution can be found by an analogous procedure of the combination of variables as was used to derive the Leveque solution (17.37). The solution, when transposed back in the dimensional variables, can be expressed as

$$T = T_0 + \frac{q_w\Delta_2}{k}\Psi_2\left(\frac{R-r}{\Delta_2}\right) \tag{17.67}$$

where the boundary layer thickness is

$$\Delta_2 = 1.536\left(\frac{z\,a}{w'}\right)^{1/3} \tag{17.68}$$

and the respective dimensionless temperature profile $\Psi_2((R-r)/\Delta_2)$ is also graphically displayed in Fig. 17.2.

While the intensity of the local heat flow through the wall is

$$q(z) \equiv k\frac{\partial T(r,z)}{\partial r}\bigg|_{r=R} = \frac{k(T_0 - T_W)}{\Delta_2} \tag{17.69}$$

then in the tube

$$\lim_{\zeta\to 0} Nu_z(\zeta) = 1.302\left(\frac{z}{dRePr}\right)^{-\frac{1}{3}} \tag{17.70}$$

Graphically the complete solution for large interval of z is shown with the asymptotes in Fig. 17.4.

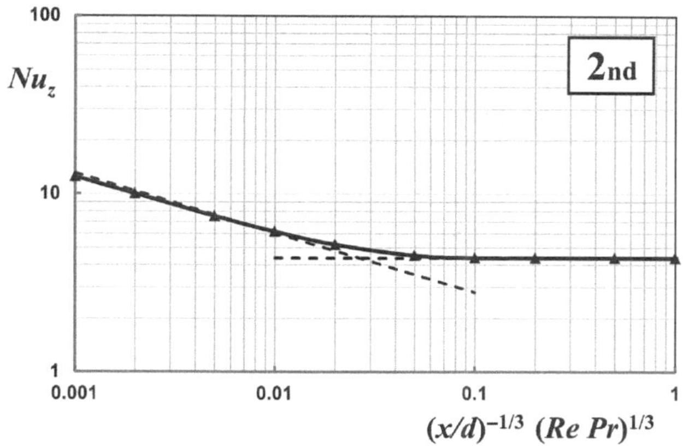

Fig. 17.4 The local Nusselt number for the constant heat flow through the wall of a circular tube and the asymptotes (17.70) and (17.65).

17.5 The generalization for more realistic boundary conditions

In the engineering practice it is difficult to realize exactly the condition of a wall temperature jump and then keeping the temperature constant (the first kind of the boundary condition) neither to maintain the heat flow constant (the second type of the boundary condition). In reality, the absolute values of the temperature difference $|T_W - T_0|$ usually somewhat grow with the growing distance z while the intensity of the heat flow $|q_W|$ somewhat drops. It could be thus expected that the value of the local Nusselt number will be somewhere between the relatively close values of the two limiting cases (Fig. 17.5).

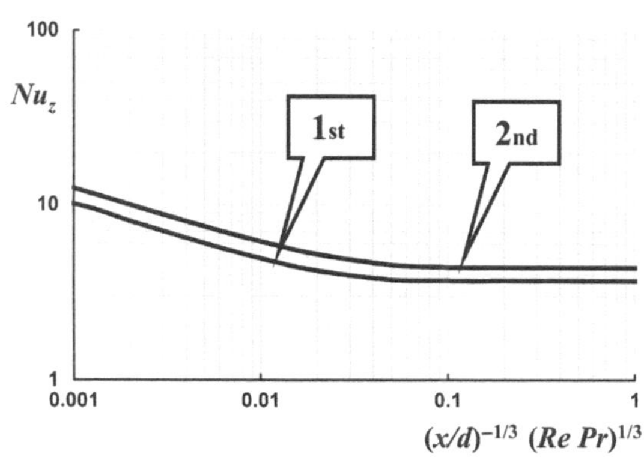

Fig. 17.5 The local Nusselt number for a circular tube at different boundary conditions.

In the engineering practice it is common to choose the Nusselt number between the estimates of the two limiting situations (17.55) and (17.70) for low values of ζ and between Eqs. (17.58) and (17.65) for high values of ζ while the borderline between the low and the high values is roughly at $\zeta \approx 0.1$, that is, $(x/d)/(RePr) \approx 0.25$. A reasonably good estimate says that at very small values of ζ the situation rather follows the second boundary condition while later, in the direction of flow, it is closer to the first condition.

The unknown temperatures $T_W(z)$ and the unknown heat flow $q_W(z)$ can be calculated back from the heat balance with a not too complicated integration of the Nusselt number.

17.6 The heat transfer in a boundary layer

In a situation when a fluid of known temperature T_0 streaming at a velocity U collides with an immobile obstacle that has a temperature T_W or if a body of temperature T_W penetrates a fluid having a temperature T_0 with a velocity U then it is useful to setup a coordinate system x, y, z the origin of which will be positioned in the point of the impact of the fluid on the body. The coordinate x would track downstream the contour of the body while y leads in the normal direction away from the surface of the body. The simplest example is the flow past a flat plate the hydrodynamic situation of which was already treated in this text. There it was shown how a velocity boundary layer, the thickness of which δ_I originates. At higher Reynolds numbers, it is indirectly proportional to the root

$$\delta_I \sim \sqrt{\frac{x\mu}{\rho U}} \tag{17.71}$$

This expression can be compared with the development of the thermal boundary layer Δ in a stationary medium where

$$\Delta_I \sim \sqrt{\frac{k}{\rho c_P} t} \tag{17.72}$$

It has already been shown that the ratio of the speeds with which the velocities and the temperatures equalize is characterized by the value of the Prandtl number

$$Pr \equiv \frac{\mu c_P}{k} \tag{17.73}$$

For ordinary liquids, particularly for those being more viscous, the Prandtl number is much higher than one, $Pr \gg 1$. (For example, for water at room temperature the Prandtl number is close to $Pr \approx 7$.) In this case, a significant portion of the heat transfer occurs only inside the velocity boundary layer. To solve the standard longitudinal flow past a flat plate, the shear rate at the wall taken from Eq. (17.25) is substituted in relationship (17.38) to obtain

$$q(x) = 0.538k(T_0 - T_W)\left(\frac{\rho c_P}{k}\right)^{1/3} \frac{\left(0.332\frac{U}{x}Re_x^{1/2}\right)^{1/2}}{\left(\int_0^x \sqrt{0.332\frac{U}{x}Re_x^{1/2}}dx\right)^{1/3}} \tag{17.74}$$

so that the local Nusselt number turns out to be

$$Nu_x \equiv \frac{h_x x}{k} = 0.339 Re_x^{\frac{1}{2}} Pr^{1/3} \quad \text{for } \Delta_I \ll \delta_I \tag{17.75}$$

Remarkable is also the heat transfer in the vicinity of the stagnation point as is the case when the flow impinges on a wall of constant temperature (Fig. 15.7). The speed of the perpendicular flow in the thermal boundary layer can be calculated from relationship (15.53) with the use of the continuity equation. The thermal boundary layer at generally higher Prandtl numbers is significantly thinner than the velocity boundary layer in which it is, after Eq. (15.56) and

$$w'(x) \equiv \frac{U_x}{\delta_I} = \frac{U_x}{0.811}\sqrt{\frac{xU_x\rho}{\mu}} \tag{17.76}$$

so that a relation comes out with the use of Eq. (17.38)

$$Nu_x = 1.227\left(\frac{xU_x\rho}{\mu}\right)^{\frac{1}{2}} Pr^{1/3} \tag{17.77}$$

It is necessary to note that the less common case of $Pr \ll 1$ (usually applies to molten metals) does not fall in this category. In this case, most of the heat transfer takes place beyond the velocity boundary layer where the velocity varies very little along the y coordinate. There is a certain similarity with the piston flow. The residence time in the zone where the heat transfer takes place will be of the order

$$t \sim \frac{L}{U} \tag{17.78}$$

so that it yields in agreement with Eq. (14.53)

$$Nu_x \sim Re_x^{\frac{1}{2}} Pr^{1/2} \quad \text{for } \Delta_I \gg \delta_I \tag{17.79}$$

The general conclusion states that for external flow it is

$$Nu_x \equiv \frac{h_x x}{k} = \sim Re_x^{\frac{1}{2}} Pr^b \tag{17.80}$$

where the exponent b is

$$b = \begin{cases} {}^1/_3 & \text{for } Pr \geq 1 \\ {}^1/_2 & \text{for } Pr \ll 1 \end{cases} \tag{17.81}$$

17.7 Heat transfer in a turbulent flow

There are two popular models to interpret the heat transfer between a wall and a turbulent fluid flow. The fundamental concept and the simplest conclusions are presented in this chapter. There is a bunch of further supplements and improvements of these conclusions but they only refine some of the theoretical logical steps of the derivation. In principle, however, they do not differ substantially from these two models.

17.7.1 The film theory

The film theory emanates from the notion that an eddy of a characteristic length x is formed where a fluid impinges on a wall having a characteristic dimension L with a velocity U. The characteristic length x becomes smaller as the flow gets more intensive, that is, as the Reynolds number gets larger. A power relation is anticipated

$$x/L \sim Re_L^{-a} \tag{17.82}$$

where $a > 0$.

It is further assumed that along the length of an eddy the transfer of heat is governed by the rules of laminar flow past a plate. Because the thickness of the velocity boundary layer on the wall of the length x that is bypassed by a fluid of the velocity U will be, according to the theory of the laminar boundary layer, at higher Reynolds numbers, given by a proportionality (15.56),

$$\delta_I \sim \sqrt{\frac{x\mu}{\rho U}} \tag{17.83}$$

It follows, according to Eq. (17.75) that

$$Nu \equiv \frac{h_x L}{k} = \frac{L}{\Delta_I} \sim \left(\frac{L}{x}\right)^{1/2} \left(\frac{LU\rho}{\mu}\right)^{1/2} \left(\frac{\mu c_P}{k}\right)^{1/3} \tag{17.84}$$

or

$$Nu_L \sim Re_L^{\frac{1+\propto}{2}} Pr^{1/3} \tag{17.85}$$

Because the impact of the fluid on the wall is fast renewed it is assumed that the Nusselt number for a turbulent flow along the stream is constant.

At lower values of the Prandtl number $Pr < 1$ (less frequent case of high heat conductivity along with a low viscosity—e.g., molten metals) is the hydrodynamic boundary layer smaller than the thermal boundary layer so that the heat convection is governed rather by the rules for a plug flow

$$\Delta_I \sim \left(\frac{kx}{\rho c_P U}\right)^{1/2} \tag{17.86}$$

According to Eq. (17.79) it leads to

$$Nu \equiv \frac{h_x L}{k} = \frac{L}{\Delta_I} \sim \left(\frac{L}{x}\right)^{1/2} \left(\frac{LU\rho}{\mu}\right)^{1/2} \left(\frac{\mu c_P}{k}\right)^{1/2} \tag{17.87}$$

or

$$Nu_L \sim Re_L^{\frac{1+\propto}{2}} Pr^{1/2} \tag{17.88}$$

The determination of the characteristic size x of an eddy is possible by the statistical processing of the time series of data taken of the local instantaneous velocities of the fluid. Such data can be obtained experimentally by performing the measurement with a transparent fluid containing dispersed microscopic tracing particles.

The Laser Doppler Anemometry (LDA) involves the most demanding and costly experimental techniques. The time run of the local velocity and its oscillation is obtained by aiming a pack of laser beams focused on a specific spot.

Particle image velocimetry makes use of the video recording of the positions of all particles in a given spatial optical cut. An analysis of the successive pictures will yield the values of the instantaneous velocities in the selected plane.

Tracking of local velocities in nontransparent fluids is, of course, more difficult.

The study of eddies in turbulent flows is, for the time being, limited to specific cases. A basic knowledge that would contribute toward the general modeling of processes of heat transfer in turbulent flows is, at present, missing.

17.7.2 The model of isotropic turbulence

The use of the model of isotropic turbulence is one of the ways how to determine independently the value of α. Kolmogorov developed the idea that the turbulent fluid is filled with eddies, the size of which determines the amount of energy dissipated by viscous friction. In such eddy, the characteristic size of which is x, is the intensity of energy dissipation epsilon (that is, the energy lost in a unit of mass) proportional to

$$\varepsilon \sim \frac{\mu}{\rho}\left(\frac{U}{x}\right)^2 \tag{17.89}$$

This notion would be useful if it were possible to determine independently the distribution of the energy dissipation ε. However, this is hardly ever possible. However, what is possible is the determination of the overall power input P. Assuming further that the dissipation is uniform it is realistic to surmise that the local intensity of dissipation is related with the specific power input P/V in the given volume

$$\varepsilon \sim \frac{P}{\rho V} \tag{17.90}$$

This concept is relatively well acceptable where the turbulent whirling is relatively uniform and where the experimental data of the power input at high Reynolds numbers are fairly reliably known. The power input can be determined from quite reliable experimental data, for example, from the power consumed by an agitator in an intensively mixed vessel or from the pressure loss in fluids flowing in pipes and fittings. By combining the well-known relationships a proportion for a pipe flow is obtained

$$x/d \sim (\Lambda Re)^{-1/2} \tag{17.91}$$

where Λ is the friction coefficient for the flow in a pipe defined by Eq. (12.6). In turbulent flows $\Lambda \approx$ const. while at low Reynolds numbers it is the product (ΛRe) that is constant.

Similarly, in an agitated vessel it will be

$$x/d \sim (PoRe)^{-1/2} \tag{17.92}$$

where Po is the dimensionless **power number** that can also be expressed as

$$Po \equiv \frac{P}{\rho N^3 d^5} \tag{17.93}$$

In turbulent flows, the Power number is constant $Po \approx$ const., while at low Reynolds numbers the expression $(PoRe) \approx$ const.

In both these cases, the turbulent character of the flow results in

$$\alpha = 1/2 \tag{17.94}$$

From here the dependence of the Nusselt number follows

$$Nu_L \sim Re_L^{\frac{3}{4}} Pr^{1/3} \tag{17.95}$$

If, however, the distance L for which the flow of the liquid is in contact with the wall is short, then there is not enough time for the eddies to renew themselves and then it is possible to assume that $x = L$. It follows from there that the Reynolds number will be raised to a lower power. Thus

$$Nu_L \sim Re_L^{\frac{1}{2}} Pr^{1/3} \tag{17.96}$$

17.7.3 The penetration theory

The penetration theory stems from the notion that the fluid persists motionless for a while at the wall and is subsequently exchanged for a fresh fluid from an isothermal space. According to the theory of the unsteady heat transfer into a half-space it will be

$$\Delta_I \sim \left(\frac{k}{\rho c_P} t\right)^{\frac{1}{2}} \tag{17.97}$$

There are heat exchangers with a wiped wall, particularly for viscous liquids (low Re_L) in which the period of wiping is given explicitly by a relationship

$$t = \frac{L}{U} \tag{17.98}$$

where U is the speed of the motion of the wiper and L is the length of the section in which the liquid remains motionless until the new passage of the wiper occurs. It then applies simply

$$Nu_L \sim Re_L^{\frac{1}{2}} Pr^{1/2} \tag{17.99}$$

It is reasonable to assume that in a turbulent flow, the shorter the time for which the eddy perseveres at the wall the more intensive is the flow thus

$$t \sim \frac{L}{U} Re_L^{-\beta} \tag{17.100}$$

where $\beta > 0$.

Then a relationship criteria can be expected

$$Nu_L \sim Re_L^{\frac{1+\beta}{2}} Pr^{1/2} \tag{17.101}$$

17.7.4 The generalization of the turbulent convection

To checkup which of the given models fits better the turbulent flow is possible only on the basis of reliable experimental data. Because the Reynolds number can be set in a wide range of variables it appears in typical empirical correlations such as

$$Nu_L = A Re_L^a Pr^b \tag{17.102}$$

Those appearing most frequently in engineering literature are

$a = 0.8$	for the turbulent flow in pipes (Dittus-Boelter correlation)
$a = 2/3$	for a mixing batch

These two values fit both models.

The Prandtl number and its power could, potentially, make it clear which of the two models fits the reality better

$b=0.4$	for the turbulent flow in pipes (Dittus-Boelter correlation),
$b=1/3$	for an agitated batch and the usual $Pr>1$.

It appears, at least for a pipe, that the realistic values are somewhere between the predictions of the film and the penetration models. To carry out experiments with a wider range of Prandtl numbers that could help in making the choice is difficult since the heat conductivities and the heat capacities lie in relatively narrow limits. Increasing of the viscosity translates in the reducing of the Reynolds number to the point where the flow becomes laminar.

17.7.5 The change of viscosity

Another problem present at heat transfer tasks is the viscosity changing from place to place due to the changing temperature. Therefore the viscosity μ_M at the mean temperature of the fluid is used in all the dimensionless numbers. In addition, this step is supplemented by adding another dimensionless number (μ_M/μ_W) that expresses the ratio of the mean viscosity to the viscosity at the wall temperature. It turns out that a suitable correlation is

$$Nu_L = ARe_L^a Pr^b \left(\frac{\mu_M}{\mu_W}\right)^c \tag{17.103}$$

in which for c the most often taken empirical value $c=0.14$ is determined experimentally for the turbulent flow in pipes (the Sieder-Tate correlation). For the laminar flow the value of c is higher. It could be up to $c=1/4$.

17.8 The Chilton-Colburn analogy

For the flow past a flat plate while $Pr \gg 1$ the heat transfer can be solved by the method of the combination of variables. The modified Leveque solution with Eq. (17.75) yields

$$Nu_x \equiv \frac{x}{\Delta_I} \approx 0.3391 Re_x^{\frac{1}{2}} Pr^{1/3} \tag{17.104}$$

Because the friction coefficient for the flow past a flat plate is after Eq. (15.46)

$$\frac{f_x}{2} Re_x \equiv \frac{x}{\delta_I} \approx 0.3322 Re_x^{\frac{1}{2}} \tag{17.105}$$

an idea occurred that an analogy applies relatively fitting between the flow and the heat transfer (under the assumption of high Re and Pr) in the form of

$$\frac{Nu_x}{Re_x Pr^{1/3}} \approx \frac{f_x}{2} \tag{17.106}$$

This relationship, known as the **Chilton-Colburn analogy** leads to quite acceptable results also for gases where the Prandtl number is close to one because in such a case the power of the Prandtl number does not manifest itself.

A similar situation occurs for a perpendicular inlet on the wall (Fig. 15.7). It has been shown (15.57) that

$$\frac{f_x}{2} = 1.233\sqrt{\frac{\mu}{xU_x\rho}} \tag{17.107}$$

and then

$$Nu_x = 1.227\frac{x}{L}\left(\frac{UL\rho}{\mu}\right)^{\frac{1}{2}} Pr^{1/3} \tag{17.108}$$

it is also

$$\frac{Nu_x}{Re_x Pr^{1/3}} = 0.995\frac{f_x}{2} \tag{17.109}$$

This analogy is also quite fitting in some more complex situations which can be documented in the case of heat transfer in a pipe with a constant wall temperature

For a **laminar developed pipe**, it applies that (12.10)

$$\frac{f}{2} = \frac{16}{Re} \tag{17.110}$$

while for the heat transfer the boundary conditions were given by relationships (17.57) and (17.63):

$$Nu = Nu_\infty = \begin{cases} 3.567 & \text{for constant wall temperature} \\ 4.364 & \text{for constant heat flow} \end{cases} \tag{17.111}$$

so that the combination yields

$$\frac{Nu}{Re\,Pr^{1/3}} \approx \left(\frac{Nu_\infty}{8Pr^{1/3}}\right)\frac{f}{2} \tag{17.112}$$

Not even here the expression in the bracket for water differ substantially from one.

For the **turbulent developed pipe flow**, only empirical relationships are known out of which the most common one is the correlation of experimental data obtained for the resistance to flow in smooth pipes (Blasius)

$$\frac{f}{2}=0.0396\,Re^{-\frac{1}{4}} \tag{17.113}$$

The favorite relationship for heat transfer at high Reynolds numbers is the empirical correlation by Ditus and Boelter.

$$Nu=0.023\,Re^{0.8}Pr^{0.4} \tag{17.114}$$

so that the combination of the two yields Boelter

$$\frac{Nu}{Re\,Pr^{1/3}}=\left(0.58\,Pr^{0.067}Re^{0.05}\right)\frac{f}{2} \tag{17.115}$$

in which the expression in the brackets is not too different from one. It is also not too dependent on the change of the conditions.

In general, it turns out that the Chilton-Colburn analogy captures the nature of flow and heat transfer relation also in a number of other situations. However, the accordance of the numerical constants is not always too appealing (e.g., for the inflow on the wall as shown in Fig. 15.7).

In this case, the Chilton-Colburn analogy is used, with varying successfulness because the transfer of mass is governed formally by the same equations as the transfer of heat. All that is needed is to exchange the Nusselt number for a Sherwood number and the Prandtl number for a Schmidt number.

17.9 Summary

The principal problem of heat convection is the heat transfer from a wall during a laminar flow of a liquid through a tube. Its solution requires a number of special functions even for so simple a set of boundary conditions as is a temperature jump on the wall. This task has, however, two important asymptotes. For a long tube, that is, for a large Peclet number, it is the exponential equalization of the entry temperature profile with the temperature of the wall in which case the Nusselt number has a constant limit. Inversely, in the case of a short contact with the wall the transfer of the heat takes place only in the close vicinity of the wall so that a very thin thermal boundary layer is formed in which, using the Leveque method, a simplified solution can be found: The Nusselt number decreases with the third root of the length of the contact with the wall. Similar solutions exist for the case of the tube with a step of heat flow through a wall or for more general boundary conditions.

The value of the Prandtl number controls the flow past a wall. For common liquids the Prandtl number tends to be higher and then the thermal boundary layer is thinner than the velocity boundary layer. The Nusselt number is then proportional to the second root of the Reynolds number and the third root of the Prandtl number. This conclusion holds for most bypassed geometries.

There are several theories of how the Nusselt number for the turbulent flows depends on the Reynolds and the Prandtl numbers but the specific constants for these theories cannot be obtained by the direct solving of the transport equations. They can be estimated from related analogies but then they have to be checked by performing an experiment.

CHAPTER EIGHTEEN

Mutually influenced transfer of momentum and heat

Contents

In the foregoing example of heat convection in laminar pipe flow, a solution of the heat convection was shown under the condition when the velocity field can be first independently determined by solving first the Navier-Stokes equation and then the Fourier-Kirchhoff equation. There are, however, a number of situations when this procedure would not be quite correct because the temperature field influences the velocity field due to the **changes in density**.

- This is the case when the velocity field is impaired by so-called **free (or natural) convection**.

Equally, the velocity field is deformed by the **changes of viscosity** due to varying temperatures.

- This is the case of the so-called **nonisoviscous** (not quite conclusively also called **nonisothermal**) **flow**.

Another interference in the simple solution of the heat transfer equations is the existence of heat sources or sinks due to the reversible or irreversible changes of enthalpy due to the motion of the fluid.

- In the case of a distinct drop of pressure in a flowing gas, the heat loss might occur due to the **adiabatic expansion** or due to the **Joule-Thomson effect** (negative or positive).
- In liquid, the more likely cause is the irreversible transformation of the mechanical energy (the **energy dissipation**) in heat.

Strictly speaking, the information about the distribution of velocities and temperatures should be obtained by simultaneously solving the

Transport and Surface Phenomena
https://doi.org/10.1016/B978-0-12-818994-8.00018-X

Navier-Stokes and Fourier-Kirchhoff equations. Nowadays, this could be done using advanced software packages yet there is still a chance to make use of some simple procedures that will be presented in the following paragraphs.

In addition, when solving a concurrent heat transfer, it is always necessary to take into account the process of **thermodiffusion** or a possible thermal coloring of a **chemical reaction**. This range of problems is so broad that it goes beyond the scope of this text and it is better to deal with it in the subject of "Chemical reactors."

18.1 Free (natural) convection

In the case of a natural convection, the flow past the heat exchanging surface is not induced by some distant source (such as a pump or an agitator, an outside pressure, or a hydrostatic influence) but it is triggered by a hydrostatic force originating close to the heat transmission surface due to the changes of density as a result of temperature variation. In this case, the knowledge of the velocity U that could be used for the formulation of the Reynolds number is missing. The velocity derived from the density change can be estimated from the hydrostatic force that acts on the liquid in the layer adhering to the wall that has the thickness Δ (see Fig. 18.1).

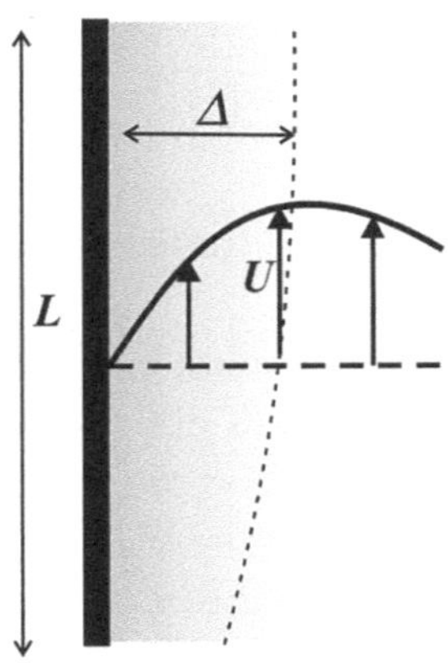

Fig. 18.1 The velocity distribution for the heating of liquid at a vertical wall (height L, width W).

If the density of the liquid adhering to the wall differs from the density in the core of the liquid then this difference generates a force of buoyancy proportional to $\Delta\rho g L W \Delta$ that will be compensated by a tangential force $\tau L W \sim \mu (U/\Delta) L W$ on the surface of the wall. A proportionality then can be expected

$$U \sim \frac{\Delta\rho g \Delta^2}{\mu} \tag{18.1}$$

The value of the Reynolds number can be estimated from known quantities in the form of

$$Re \sim \frac{U L \rho}{\mu} = \frac{\Delta\rho \rho g L \Delta^2}{\mu^2} = \frac{Gr}{Nu^2} \tag{18.2}$$

in which a Grashof number was introduced by

$$Gr \equiv \frac{\Delta\rho \rho g L^3}{\mu^2} \equiv \frac{\beta \Delta T \rho^2 g L^3}{\mu^2} \tag{18.3}$$

where β is the coefficient of extensibility

$$\beta \equiv \frac{1}{\rho} \frac{d\rho}{dT} \tag{18.4}$$

It is expected that the Nusselt number will depend on the Reynolds and Prandtl numbers by a power-law type of relationship, a general expression follows from

$$Nu \sim Gr^a Pr^b \tag{18.5}$$

where the exponent a will be smaller than the power of the Reynolds number for the forced flow. In most cases, the exponents are nearly equal $b \approx a$, so that the frequently used relationship is

$$Nu = A \left(Gr\, Pr \right)^a \tag{18.6}$$

For slow laminar free flow at $Gr\,Pr < 10^8$, the exponent a comes out to be 1/4 while for higher values of $Gr\,Pr > 10^8$, the a comes out 1/3. The value of the constant Lambda depends on the specific geometry of the system. It usually is around 0.5.

The origin of the **dissipative structures** can be clearly demonstrated using the case of free convection. Let us imagine a liquid between two parallel plates (Fig. 18.2). The lower will be warmer and the upper cooler. It follows from the symmetry of this setup that a homogeneous temperature field will be established in which the heat will be transported by mere

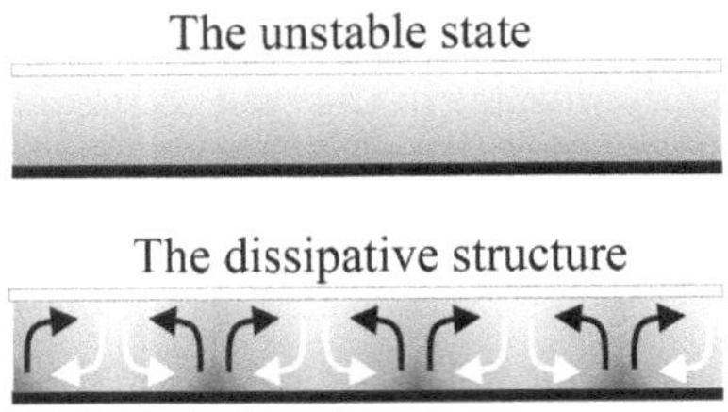

Fig. 18.2 Heating of a liquid layer from the bottom side.

conduction. Any miniscule disturbance, however, will cause the heavier liquid to descend in some places while in other places the lighter liquid will rise (the Rayleigh-Bernard instability). It is remarkable that the flow, initially quite chaotic, will organize itself in a relatively stable structure with individual hexagonal ground cells of almost identical magnitude in which the fluid circulates.

The density of water increases with increasing temperature when the temperature of water under the ice reaches 4°C at the bottom. Then the natural convection ceases and further cooling of the water occurs only due to conduction which is very slow.

The free convection of mass transport is also possible if the density of the fluid changes significantly in the process. A good example of this process is the dissolution of crystals leading to a denser solution.

A significant example of this process in nature happens when the sea freezes over. The ice has a lesser content of salt than water so that the remaining saltwater sinks to the sea floor and thus creates specific sea currents at the ocean floor.

18.2 Nonisoviscous ("nonisothermal") flow

If the heat transmitted to flowing fluid results in a temperature change then such a change will most likely translate in a change of viscosity. The dependence of viscosity on temperature can be well approximated by a relationship

$$\mu \approx \mu_0 \exp\left[A\left(\frac{1}{T}-\frac{1}{T_0}\right)\right] \tag{18.7}$$

where μ_0 is the viscosity at the temperature T_0. Frequently, a simpler approximate relationship will suffice

$$\mu \approx \mu_0 \exp\left(\frac{T - T_0}{T_\mu}\right) \tag{18.8}$$

Examples of the values of T_μ and μ_0 for the temperature $T_0 = 50°C$ are given in Table 18.1.

Table 18.1 Temperature dependence of viscosity.

	$T_0 = 50°C$ μ_0 (mPa s)	T_μ (K)
Steam	0.0100	500
Air	0.0196	260
Methanol	0.396	−82
Water	0.561	−56
Gear oil	95	−18
Glycerine	180	−19
Honey 85%	1500	−11
Silicon oil	19,000	−65

The viscosity of gases increases with increasing temperature, while the viscosity of liquids drops. The variable T_μ expresses the temperature difference for which the viscosity will increase by 2.72 times. It is evident that the changes of viscosity with increasing temperature will be most manifested for highly viscous liquids (an exception are the silicon oils). In the laminar flow of liquids chilled externally through a wall of a pipe, the parabolic profile can be deformed into a shape shown in Fig. 18.3A when heating it acquires a shape shown in Fig. 18.3B.

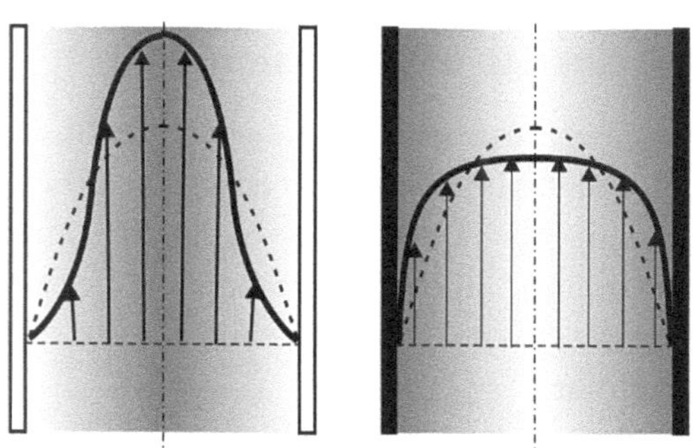

(A) $\mu_W > \mu_M$ (B) $\mu_W < \mu_M$

Fig. 18.3 Examples of velocity profiles in a pipe for a laminar flow. The dotted profile corresponds to $\mu = \text{const}$.

Keeping in mind that, after Leveque, the coefficient of heat transport is proportional to the third root of the velocity gradient at the wall, the Nusselt number will be lower than it would correspond to a parabolic profile when the liquid is cooled while it is higher when the liquid is warmed (the other way round for gases). The respective solution can be obtained only numerically. If the direction of the flow is known, then it is possible to assume that the changes of the velocity field are somewhat slower than the changes in the temperature field. It is thus possible, using a step-by-step procedure, starting from the actual velocity profile to determine the following temperature change and from it to set the next velocity profile.

In the case of gases for which the velocity and the temperature profiles change in parallel (due to the Prandtl number being close to one), the velocity profile should develop conversely than for liquids. However, the turbulent flow is more common in the case of gases so that the transverse homogenization of temperatures is more common.

On the basis of experimental data, Sieder and Tate (1936) suggested an approximate formula (18.9) to compare the real Nusselt number $Nu_{nonisothermal}$ with an ideal case $Nu_{isothermal}$ making use of the ratio of the viscosity at the wall μ_W to the viscosity μ_M at the mean temperature of the fluid

$$Nu_{nonisothermal} = Nu_{isothermal}\left(\frac{\mu_M}{\mu_W}\right)^c \tag{18.9}$$

where the exponent c is usually taken $c=0.14$ for any turbulent flow. In the case of the laminar flow, the influence of viscosity is slightly underestimated.

18.3 The influence of expansion

If a strong gas pressure drop occurs in equipment (usually due to throttling in an element with high resistance to flow), then this could lead to the exertion of mechanical work at the expense of the inner energy. This process would result in the cooling of the whole volume. In the case of liquids, since their volume can hardly change, the mechanical work is slight, and this phenomenon is insignificant.

Apart from this phenomenon that can be realized with almost ideal gases, there is another way of conversion of mechanical work in the heat, the Joule-Thompson effect. During an adiabatic expansion, the temperature will be subject to a change

$$\left(\frac{\partial T}{\partial P}\right)_H = \mu_{JT} \tag{18.10}$$

The symbol μ_{JT} is known as the Joule-Thompson coefficient. This coefficient is positive at low temperatures below the so-called inversion temperature, but it is negative above it.

For most real gases at usual temperatures, the expansion results in cooling down. Inversely, the compression in warming up, particularly if accompanied by liquefaction. This phenomenon is utilized in heat pumps such as refrigerators and air conditioners.

The inversion temperature for the main components of air is around 300°C. At room temperature, the μ_{JT} is about 0.5 K/bar. The inversion temperature for hydrogen is −80°C. This makes potential escapes of this gas at current temperatures from a pressurized space dangerous because, under this condition, the hydrogen's $\mu_{JT} = -0.05$ K/bar and the escaping hydrogen warms up in the process. As a result, the risk of fire and explosion increases.

The Joule-Thompson effect is also exploited for the liquefaction of air.

When solving the problems of heat transfer with the influence of expansion, it is common to add to the equation one more term, a source term that is almost constant across the whole cross section. Certain solutions might also be then accessible analytically.

18.4 The dissipation of mechanical energy

Like the friction of solid bodies, the flow of viscous fluids leads to irreversible losses, **dissipation**, of mechanical energy that converts to heat. In a laminar flow, the loss of mechanical energy in a unit volume is given by the product of the tangential stress tau and the shear rate gamma. The corresponding rate of heating up is

$$\frac{dT}{dt} = \frac{\varepsilon}{\rho c_P} = \frac{\tau\gamma}{\rho c_P} = \frac{\mu\gamma^2}{\rho c_P} \tag{18.11}$$

In a flow tube, the pressure loss in a flowing fluid is Δp and the mean mixed temperature will rise due to viscous friction by ΔT_M whereby it is

$$\Delta T_M = \frac{\Delta p}{\rho c_P} \tag{18.12}$$

This phenomenon is particularly pronounced in very viscous liquids. It is utilized in the processing of polymers where the dissipation heat is one of the ways of supplying the heat to the batch. Its key advantage is a certain self-controlling effect because warming decreases the viscosity and thus the amount of the heat produced. As a result, the risk of damaging the polymer is reduced compared with the heating through a hot wall.

For the solution of problems involving energy dissipation, it is usual to proceed as in solving nonisoviscous flows that is by a step-by-step adjusting of the velocity profile and the determination of the dissipative heat source. Only in an idealized case when the temperature variation is passed over the problem can be solved analytically relatively easily.

18.5 Summary

In the preceding chapter, those cases were discussed when the heat convection was influenced by a known flow of a fluid. In reality, it is the flow that is affected by heat transmission because both the density and the viscosity of the fluid are fundamentally affected by temperature.

The density variation is manifested by the sinking of the denser fluid down and the consequent replacement of the lighter layers upward. This results in spontaneous streaming—the natural convection. Its significance is indicated by the magnitude of the Grashof number. The complete solution to the problem of flow with simultaneous heat transfer is possible with today's available mathematical means. It is the foundation of the meteorological models. In the engineering practice, it is usually satisfactory to work with rough estimates.

The viscosity variation with the temperature will also affect the forced convection since the velocity gradient dropped in the space of higher viscosity while in the region of lower viscosity the flow and the heat transfer will be enhanced. The ratio of the mean temperature and the wall temperature is used to estimate the increase of the Nusselt number.

In the case of gases, the heat balance will be affected by the flow when large pressure changes are manifested by changes of the enthalpy.

The temperature field can also be affected by the flow of very viscous liquids when a significant portion of the mechanical energy is dissipated in heat.

References

Sieder, E.N., Tate, G.E., 1936. Ind. Eng. Chem. Res. 28, 1429–1435.

PART THREE

Mathematical methods for solving the transfers

CHAPTER NINETEEN

Vectors and tensors

Contents

19.1 Vectors

In this chapter, vectors are not being dealt with in the algebraic sense (an arranged matrix of elements of a set) but as the ordinarily used geometrical vectors in a three-dimensional (3D) Euclidean space. Most of the rules will be shown formalized in the rectangular Cartesian coordinates. However, vector is an object completely independent of the choice of the coordinates and it can be simply transposed from one system to another.

In a handwritten text, a vector is denoted by an arrow above the symbol, $\vec{a}$, In a printed text, a vector is printed bold, $\boldsymbol{a}$. This rule will also be applied in this chapter.

Geometrically, a vector can be characterized by an absolute magnitude and a direction. A vector has its beginning and an end and this mutually bound couple is freely movable in the space (see Fig. 19.1).

As a rule, a vector $\boldsymbol{a}$ in the Cartesian coordinate system x, y, z is described by a triplet of scalars a_x, a_y, a_z, whereby a_x is the oriented length of the projection of the vector into the x-axis, etc. If a unit vector in the Cartesian coordinate system in the x-direction is denoted as $\boldsymbol{i}_x$, in the y-direction $\boldsymbol{i}_y$, and in the $\boldsymbol{i}_x$-direction $\boldsymbol{i}_z$, then it can be expressed

Transport and Surface Phenomena
https://doi.org/10.1016/B978-0-12-818994-8.00019-1

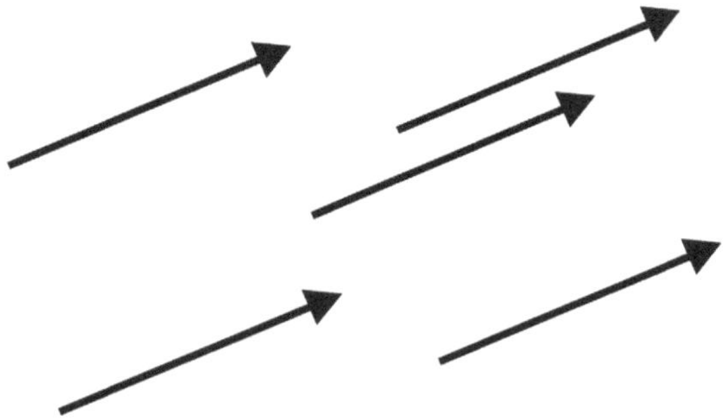

Fig. 19.1 Identical vectors.

$$a = a_x i_x + a_y i_y + a_z i_z \tag{19.1}$$

whereby the scalars a_x, a_y, a_z can be positive as well as negative depending on the orientation of the vector.

19.1.1 Selected operations with vectors

The operations with vectors are subject to several fixed mathematical rules, for example, the sum of vectors ***a*** and ***b*** is a vector ***s*** for which it applies simultaneously

$$s_x = a_x + b_x, \quad s_y = a_y + b_y, \quad s_z = a_z + b_z \tag{19.2}$$

This can be written as

$$s = a + b \tag{19.3}$$

As is obvious from Fig. 19.2, the commutative law applies:

$$\begin{aligned} s &= a + b \equiv b + a \\ a &= s - b \\ b &= s - a \end{aligned} \tag{19.4}$$

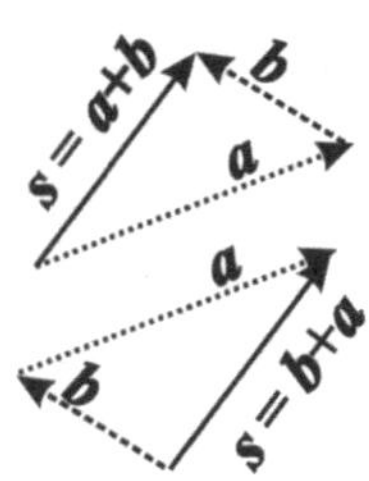

Fig. 19.2 The commutative law.

as well as the associative law:

$$(a+b)+c \equiv a+(b+c) \tag{19.5}$$

and the repetitive summation can be understood as a multiplication by an scalar, where the distributive law applies

$$k(a+b) \equiv ka+kb \tag{19.6}$$

Geometrical vectors are suitable for the description of processes taking place in a plane or in space. They are particularly useful for the location of a point in a chosen coordinate system (**Radius vector**). This is the vector $\boldsymbol{r}$ that connects the origin of the coordinate system with the given point. Its absolute value is expressed by the distance of the given point from the origin.

$$|\boldsymbol{r}| = \sqrt{r_x^2 + r_y^2 + r_z^2} \tag{19.7}$$

19.1.2 Derivative with respect to a vector

Quantities that are a function of a radius vector can be differentiated whereby it is useful for the shortening of the written expression to introduce the **Hamilton operator "nabla"** (alternatively also $\boldsymbol{\nabla}$) that makes sense only when applied on a variable. In the Cartesian coordinate system, it is defined as

$$\boldsymbol{\nabla} \equiv \boldsymbol{i}_x \frac{\partial}{\partial x} + \boldsymbol{i}_y \frac{\partial}{\partial y} + \boldsymbol{i}_z \frac{\partial}{\partial z} \tag{19.8}$$

When applied on a scalar s, it results in a vector termed the **gradient of a scalar**

$$\text{grad } s \equiv \boldsymbol{\nabla} s = \boldsymbol{i}_x \frac{\partial s}{\partial x} + \boldsymbol{i}_y \frac{\partial s}{\partial y} + \boldsymbol{i}_z \frac{\partial s}{\partial z} \tag{19.9}$$

Physically, the gradient of a scalar characterizes the direction and the magnitude of the largest growth of the scalar in the vicinity of the point.

19.2 The derivative of a vector

The case when a vector is a function of a spatial variable $\boldsymbol{a}(\boldsymbol{r})$ is fairly common. This case is called a vectorial field. To this function, it is also possible to determine the gradient of the vector $\boldsymbol{a}$. In general, it consists from the derivatives with respect to individual coordinates

$$\frac{\mathrm{d}\boldsymbol{a}}{\mathrm{d}t} \equiv \lim_{\Delta t \to 0} \frac{\boldsymbol{a}(t+\Delta) - \boldsymbol{a}(t)}{\Delta t} \tag{19.10}$$

19.2.1 The derivative of a vector field

The case when a vector is a function of a spatial variable $\boldsymbol{a}(\boldsymbol{r})$ is fairly common. This case is called a **vectorial field**. To this function, it is also possible to determine the gradient of the vector $\boldsymbol{a}$. In general, it consists of the derivatives with respect to individual coordinates

$$\begin{aligned}\text{grad}\,\boldsymbol{a} &\equiv \nabla\boldsymbol{a}\\ &= \boldsymbol{i}_x\frac{\partial a_x}{\partial x}\boldsymbol{i}_x + \boldsymbol{i}_y\frac{\partial a_x}{\partial y}\boldsymbol{i}_x + \boldsymbol{i}_z\frac{\partial a_x}{\partial z}\boldsymbol{i}_x + \boldsymbol{i}_x\frac{\partial a_y}{\partial x}\boldsymbol{i}_y + \boldsymbol{i}_y\frac{\partial a_y}{\partial y}\boldsymbol{i}_y + \boldsymbol{i}_z\frac{\partial a_y}{\partial z}\boldsymbol{i}_y + \boldsymbol{i}_x\frac{\partial a_z}{\partial x}\boldsymbol{i}_z\\ &\quad + \boldsymbol{i}_y\frac{\partial a_z}{\partial y}\boldsymbol{i}_z + \boldsymbol{i}_z\frac{\partial a_z}{\partial z}\boldsymbol{i}_z\end{aligned} \tag{19.11}$$

It consists of nine scalar components, each of which has two directions so that it forms a **tensor** (see Section 19.3).

19.2.2 More special derivatives of a vector field

In a number of cases, it is not necessary to have a complete derivative. Instead, it is more convenient to define other variables. Particularly useful is so-called **vector divergence** that is a scalar product

$$\text{div}\,\boldsymbol{a} \equiv \nabla\cdot\boldsymbol{a} = \frac{\partial a_x}{\partial x} + \frac{\partial a_y}{\partial y} + \frac{\partial a_z}{\partial z} \tag{19.12}$$

According to the Gauss-Ostrogradsky clause, the divergence is linked to the outflow of a vector from a differential volume ΔV through a circumferential area ΔS.

$$\text{div}\,\boldsymbol{a} = \lim_{\Delta V \to 0} \frac{\oiint_{\Delta S}\boldsymbol{a}\,\mathrm{d}S}{\Delta V} \tag{19.13}$$

The outflow of mass from a unit volume was discussed while dealing with the equation of continuity (Section 9.1).

Another interesting differential object is the **vector rotation**

$$\mathbf{rot}\,\boldsymbol{a} \equiv \left(\frac{\partial a_z}{\partial y} - \frac{\partial a_y}{\partial z}\right)\boldsymbol{i}_x + \left(\frac{\partial a_x}{\partial z} - \frac{\partial a_z}{\partial x}\right)\boldsymbol{i}_y + \left(\frac{\partial a_y}{\partial y} - \frac{\partial a_z}{\partial y}\right)\boldsymbol{i}_z \tag{19.14}$$

For a rotating body, there is a **rot** $\boldsymbol{u}$ vector assigned to the local velocity vector $\boldsymbol{u}$ that has the direction of the axis of the rotation and its magnitude is equal to the angular velocity.

19.3 Tensors

Cartesian tensor (second order) is a mathematical object formed by a matrix of nine scalar elements. In a handwritten text, a tensor is denoted by two arrows above the symbol $\vec{\vec{T}}$ in a printed text it is just printed bold.

$$\boldsymbol{T} \equiv \begin{Vmatrix} T_{xx} & T_{xy} & T_{xz} \\ T_{yx} & T_{yy} & T_{yz} \\ T_{zx} & T_{zy} & T_{zz} \end{Vmatrix} \tag{19.15}$$

In a geometrical meaning, this tensor can be expressed as a sum of

$$\begin{aligned} \boldsymbol{T} = {} & \boldsymbol{i_x} T_{xx} \boldsymbol{i_x} + \boldsymbol{i_x} T_{xy} \boldsymbol{i_y} + \boldsymbol{i_x} T_{xz} \boldsymbol{i_z} + \boldsymbol{i_y} T_{yx} \boldsymbol{i_x} + \boldsymbol{i_y} T_{yy} \boldsymbol{i_y} + \boldsymbol{i_y} T_{yz} \boldsymbol{i_z} \\ & + \boldsymbol{i_z} T_{zx} \boldsymbol{i_x} + \boldsymbol{i_z} T_{zy} \boldsymbol{i_y} + \boldsymbol{i_z} T_{zz} \boldsymbol{i_z} \end{aligned} \tag{19.16}$$

A tensor is thus an object having two directions. The need to introduce tensors into mathematics was inspired by the mechanics in order to be able to express a tension that is induced on each oriented area (vector $\boldsymbol{S}$) by an oriented force (vector $\boldsymbol{F}$). In this case, the assignment is given by a scalar product

$$\boldsymbol{F} = \boldsymbol{\sigma} \cdot \boldsymbol{S} \tag{19.17}$$

that, in a shorter form, captures the relations between the elements of the vectors and the tensors.

$$\begin{aligned} F_x &= \sigma_{xx} S_x + \sigma_{xy} S_y + \sigma_{xz} S_z \\ F_y &= \sigma_{yx} S_x + \sigma_{yy} S_y + \sigma_{yz} S_z \\ F_x &= \sigma_{zx} S_x + \sigma_{zy} S_y + \sigma_{zz} S_z \end{aligned}$$

For further mathematical operations, it is useful to define special tensors. **Diagonal tensor** has all elements nonzero except those in the diagonal. A special diagonal tensor is the unit tensor $\boldsymbol{\delta}$ (also known as the Kronecker delta). It is given by the matrix

$$\boldsymbol{\delta} \equiv \begin{Vmatrix} 1 & 0 & 0 \\ 0 & 1 & 0 \\ 0 & 0 & 1 \end{Vmatrix} \tag{19.18}$$

To each tensor $\boldsymbol{T}$, there could be a **transpose tensor** $\boldsymbol{T}^T$ assigned in which the rows and the columns are interchanged.

$$\left(T^T\right)_{mn} = T_{nm} \tag{19.19}$$

For a **symmetric tensor**, it applies

$$T_{mn} = T_{nm} \tag{19.20}$$

The symmetric tensor has, therefore, only six independent elements.

An **antisymmetric tensor**

$$T_{mn} = -T_{nm} \tag{19.21}$$

has only three independent elements.

A **dyad** of vectors may also be considered to be a tensor. It is usually denoted by a sign of a dyadic product $\otimes$

$$\boldsymbol{T} = \boldsymbol{a} \otimes \boldsymbol{b} \quad \text{and the tensor elements are} \quad T_{mn} = a_m b_n \tag{19.22}$$

19.3.1 Selected operations with tensors

The sum of tensor $\boldsymbol{T}$ and $\boldsymbol{U}$ is again a tensor whereby a following rule applies:

$$W_{mn} = T_{mn} + U_{mn} \tag{19.23}$$

Each tensor $\boldsymbol{T}$ can be split into a sum of a tensor symmetric $\boldsymbol{S}$ and an antisymmetric $\boldsymbol{A}$

$$\boldsymbol{T} = \boldsymbol{S} + \boldsymbol{A} \tag{19.24}$$

where

$$\boldsymbol{S} = \frac{1}{2}\left(\boldsymbol{T} + \boldsymbol{T}^T\right) \text{a} \boldsymbol{A} = \frac{1}{2}\left(\boldsymbol{T} - \boldsymbol{T}^T\right) \tag{19.25}$$

This will be in the component form

$$S_{mn} = \frac{1}{2}\left(T_{mn} + T_{nm}\right) = S_{nm} \tag{19.26}$$

$$A_{mn} = \frac{1}{2}\left(T_{mn} - T_{nm}\right) = -A_{nm} \tag{19.27}$$

The product $\boldsymbol{W}$ of a tensor $\boldsymbol{T}$ and a scalar c is tensor $\boldsymbol{W} = c\boldsymbol{T}$ in that

$$W_{mn} = c\,T_{mn} \tag{19.28}$$

The scalar product of the tensor $\boldsymbol{T}$ and a vector $\boldsymbol{a}$ from the left is a tensor $\boldsymbol{W}= \boldsymbol{a} \cdot \boldsymbol{T}$ where

$$W_{mn} = \sum_q a_{mq} T_{qn} \tag{19.29}$$

The scalar product of the tensor $\boldsymbol{T}$ and a vector $\boldsymbol{a}$ from the right is a tensor $\boldsymbol{Y}= \boldsymbol{T} \cdot \boldsymbol{a}$ where

$$Y_{mn} = \sum_q T_{mq} a_{qn} \tag{19.30}$$

Here the commutative law does not apply.

In a similar fashion, it is possible to act on a tensor $\boldsymbol{T}$ by an operator $\boldsymbol{\nabla}$ from the right as well as from the left where for

$$\boldsymbol{W} = \boldsymbol{\nabla} \cdot \boldsymbol{T} \quad \text{the tensor elements are} \quad W_{mn} = \frac{\partial T_{mn}}{\partial m} \tag{19.31}$$

and for

$$\boldsymbol{Y} = \boldsymbol{T} \cdot \boldsymbol{\nabla} \quad \text{the tensor elements are} \quad Y_{mn} = \frac{\partial T_{mn}}{\partial n} \tag{19.32}$$

Further, it is also possible to introduce a product of tensors. Particularly significant is the **double dot scalar product** in which first act on each other the inside indices and then the outside ones. This product is marked by a double dot.

$$\boldsymbol{T} : \boldsymbol{W} \equiv \sum\nolimits_{m=1}^{3} \sum\nolimits_{n=1}^{3} T_{mn} W_{nm} \tag{19.33}$$

19.3.2 Scalar invariants of a tensor

In a similar fashion as the scalar characteristics of a vector is its absolute value so are three independent scalar quantities needed to characterize a tensor $\boldsymbol{T}$. The sum of its diagonal components is usually taken as the first invariant

$$I_1 = \sum\nolimits_{m=1}^{3} T_{mm} \tag{19.34}$$

The second invariant is taken to be the double scalar product

$$I_2 = \boldsymbol{T} : \boldsymbol{T} \tag{19.35}$$

and the determinant

$$I_3 = \begin{vmatrix} T_{xx} & T_{xy} & T_{xz} \\ T_{yx} & T_{yy} & T_{yz} \\ T_{zx} & T_{zy} & T_{zz} \end{vmatrix} \tag{19.36}$$

is taken as the third invariant.

Scalar functions of a tensor can be expressed in terms of this trio.

CHAPTER TWENTY

Some ordinary differential equations

Contents

Since this text deals only with continuous functions in well-defined coordinate systems it is not necessary to handle some of the general requirements that accompany precise mathematical expressions. Instead, the focus will be only on some useful notes.

The general form of a differential equation of the first order can be formally expressed as

$$F\left(\frac{\mathrm{d}y}{\mathrm{d}x}, y, x\right) = 0 \tag{20.1}$$

and of the second order as

$$F\left(\frac{\mathrm{d}^2 y}{\mathrm{d}x^2}, \frac{\mathrm{d}y}{\mathrm{d}x}, y, x\right) = 0 \tag{20.2}$$

Equations of higher order do not usually occur in the description of transport processes.

Transport and Surface Phenomena
https://doi.org/10.1016/B978-0-12-818994-8.00020-8

20.1 Equations of the first order

20.1.1 The general equation of the first order

From the general equation

$$F\left(\frac{dy}{dx}, y, x\right) = 0 \tag{20.3}$$

can be obtained by ordinary arithmetic operations

$$\frac{dy}{dx} = f(x, y) \tag{20.4}$$

No general solution of this equation is available. There are, however, software means available also in this case that make it possible to solve the function $y(x)$ numerically if at least one initial condition is given. This could be

First: the knowledge of the function $y_0(x_0)$ appropriate to the argument x_0 (the Dirichlet condition),

Second: the knowledge of the derivative $dy/dx|_{x=x_0}$ appropriate to the argument x_0 (the Neumann condition),

Third: less often the knowledge of the linear relationship between $y_0(x_0)$ and $dy/dx|_{x=x_0}$ (Newton's condition).

Today, it is not necessary to program the whole procedure since numerical procedures are available that will perform the whole course of action without the submitter having to develop and test algorithms and strategies securing required accuracy.

20.1.2 Variable separation

Fairly often a relationship has been met in this text that can be expressed by an equation

$$\frac{dy}{dx} = f(x)g(y) \tag{20.5}$$

In it the variables can be separated into groups

$$\frac{dy}{g(y)} = f(x)\,dx \tag{20.6}$$

and the result is obtained by integration

$$\int_{y_0}^{y} \frac{dy}{g(y)} = \int_{x_0}^{x} f(x)\, dx \tag{20.7}$$

for which only one boundary condition is needed (usually that is $y = y_0$ for $x = x_0$).

A rigorous mathematician would use some other dummy variables rather than x and y behind the integration sign because he would reserve these for the upper limit.

Fairly often the integrals of this type result in primitive functions. In other cases, the integration must be performed numerically and also here software means are normally available.

It is though sometimes interesting to expand the function in the integrand into a series the terms which have simple analytical solutions; in this way it is possible to obtain an expression for the integral in some asymptotic cases.

20.2 Second-order equations

20.2.1 A general second-order equation

For a general equation

$$F\left(\frac{d^2y}{dx^2}, \frac{dy}{dx}, y, x\right) = 0 \tag{20.8}$$

no general solution exists.

20.2.2 Order reduction

The order in simpler equations such as

$$F\left(\frac{d^2y}{dx^2}, \frac{dy}{dx}, x\right) = 0 \tag{20.9}$$

can be reduced by a substitution

$$z(x) = \frac{dy}{dx}, \tag{20.10}$$

to a known form

$$F\left(\frac{dz}{dx}, z, x\right) = 0 \tag{20.11}$$

This equation can be solved by a procedure indicated in the preceding paragraph so what is left is only the integration of relation (20.10).

A similar procedure can be adopted for solving the equation

$$F\left(\frac{\mathrm{d}^2y}{\mathrm{d}x^2}, \frac{\mathrm{d}y}{\mathrm{d}x}, y\right) = 0 \tag{20.12}$$

The order of the equation can be reduced by using

$$\frac{\mathrm{d}x}{\mathrm{d}y} = 1/\left(\frac{dy}{dx}\right) \tag{20.13}$$

and

$$\frac{\mathrm{d}^2x}{\mathrm{d}y^2} = \frac{-\dfrac{\mathrm{d}^2y}{\mathrm{d}y^2}}{\left(\dfrac{\mathrm{d}y}{\mathrm{d}x}\right)^3} \tag{20.14}$$

With the help of this adjustment an equation is obtained

$$F\left(\frac{\mathrm{d}^2x}{\mathrm{d}y^2}, \frac{\mathrm{d}x}{\mathrm{d}y}, y\right) = 0 \tag{20.15}$$

The following procedure is analogous to the order reduction of Eq. (20.11).

20.2.3 Linear equation of the second order

The discussion of how to solve a general linear equation of the second order

$$a_2(x)\frac{\mathrm{d}^2y}{\mathrm{d}x^2} + a_1(x)\frac{\mathrm{d}y}{\mathrm{d}x} + a_0(x)y = g(x) \tag{20.16}$$

can be found in mathematical handbooks. Somewhat simpler is the case when the functions a_i are replaced by constants and when the function $g(x)$ is zero or if it has some specific shape.

Specimens of such more easily solvable equations were shown in the chapter. To solve more complex equations it is necessary to turn to numerical solution.

20.2.4 Nonlinear equations

Nonlinear equations are those containing products or powers or some other combination of functions. Analytical solution of such equations is impassable and the numerical procedure is a necessity. One of the key problems of nonlinear equations is the instability of their solution due to their nonlinearity,

An example is the Navier-Stokes equation for higher Reynolds numbers where the nonlinearity is due to the product of the velocity with its derivative on the left-hand side of the equation. The turbulence is thus mathematically justified chaos due to the nonlinearity of the equation.

20.3 Numerical solution

Numerical solution does not lead to a universal result valid generally for any initial or boundary conditions but it is just a case study valid for only one task.

20.3.1 Cauchy's (initial) problem

If only the initial conditions are given (Cauchy's task) then the solution progresses from the beginning step by step always estimating the value of the function in the next point. If this is done manually then some simple approximation for the short interval must be assumed. In the simplest case, this could be based on the lowest members of the Taylor expansion. Computer-based numerical solutions today choose more complex algorithms that test the variation of the function in several points of the interval and from these suggest the next most fitting value (a very popular is the method named after Runge and Kutta). In most cases, the algorithms are already built in the software packages and the user does not need to worry about their structure.

20.3.2 Boundary problem

A somewhat more complex case is the equation of the second order for which the roots in two different points of the arguments are known. In principle, each of them can be formulated as Dirichlet's (for example, by the determination of the value of the function $y(x_1)$ or $y(x_2)$), Neumann's (for example, by the determination of the value of $\mathrm{d}y/\mathrm{d}x|_{x=x_1}$ or of $\mathrm{d}y/\mathrm{d}x|_{x=x_2}$) condition, or by some other relationship of these values. Basically, there are two possible methods of solution as follows:

- To a given point at the boundary another point is guessed so as to have simultaneously the value of the function as well as its derivative. The calculation then proceeds from this point to the other. A disagreement then will appear in the second point between the calculated and the given boundary condition. Having this disagreement in mind an attempt must be made to correct the guessed value in the initial point using the usual interpolation techniques. Sometimes, the new better guess follows from the very nature of the equation.

- Several functions are chosen that have approximately the same shape as the expected outcome. Some optimalization criterion must be chosen to asses which of the chosen functions fits best. This could be, for example, the minimalization of the entropy production (Section 12.5).
- Another effective method could be to select a series of functions and then optimize their coefficients. The calculus of variations could be used to this end.

20.3.3 Calculus of variations

Calculus of variations deals with seeking an extremal of a functional. A **functional** is a function $S(f)$ that expresses dependence not only on the argument x but also on the function f. It depends also not only on the function $f(x)$ but also on its derivative. Formally, it is written as

$$S(f) = \int_a^b F\left(x, f(x), \frac{df}{dx}\right) \mathrm{d}x \tag{20.17}$$

The **extremal of a functional** is a situation when the functional at given boundary conditions bound to points a and b reaches the local extreme. In such a case the Euler-Lagrange equation of the calculus of variations applies

$$F\left(x, f(x), \frac{df}{dx}\right) - \frac{\mathrm{d}}{\mathrm{d}x} F\left(x, f(x), \frac{df}{dx}\right) = 0 \tag{20.18}$$

Using this method it is possible to transform the search for an extremal in the solution of a differential equation.

For example, searching for the shape of a hanging mass rope (so-called catenary) or searching for the minimal production of entropy.

Otherwise it is possible to transform the solution of a differential equation to the search for an extremal. Then it is possible to use, for example, the Ritz-Galerkin method testing of various functions or their expansions in series for the best approximation of the equation.

The problem can be expanded also to include the functional with multiple variables and with higher derivatives. The results are somewhat less transparent formulas but the methods of the calculus of variations can be used also in this case. It is advantageous if the region of integration has a simple shape.

In this way it is possible to calculate relatively accurately the velocity profile in a conduit of a rectangular cross section (Section 12.5).

CHAPTER TWENTY-ONE

Some partial differential equations

Contents

Mathematical interpretation of natural events often leads to partial differential equations, the solution of which is rarely trivial.

Initially, it is advantageous to strip the equation to be solved of unnecessary variables, usually by transforming it into a dimensionless form. It is also convenient to normalize them so that the boundary conditions are set to zeros or ones. Equations processed in such a way, stripped of any links to the original physical or chemical base, can then be given to mathematicians and computer specialists for final processing. These systems are often called the **equations of mathematical physics**. Most often these are linear equations of the elliptic kind taking the form

$$a(x, y)\frac{\partial^2 F}{\partial x^2} + b(x, y)\frac{\partial^2 F}{\partial x \partial y} + c(x, y)\frac{\partial^2 F}{\partial y^2} + d(x, y)\frac{\partial F}{\partial x} + e(x, y)\frac{\partial F}{\partial y} + f(x, y)F + g(x, y) = 0 \tag{21.1}$$

One of the simplest in this category is the equation of the steady plane heat conduction (13.10)

$$\frac{\partial^2 T}{\partial x^2} + \frac{\partial^2 T}{\partial y^2} = 0 \tag{21.2}$$

On the other hand, the Navier-Stokes equation is not linear because it includes product of speed and its derivative. Because of that, it might result in chaotic multiple solution (turbulence).

Transport and Surface Phenomena
https://doi.org/10.1016/B978-0-12-818994-8.00021-X

Substantially simpler is the solution of the **linear equation of the parabolic type**

$$a(x, y)\frac{\partial^2 F}{\partial x^2} + d(x, y)\frac{\partial F}{\partial x} + e(x, y)\frac{\partial F}{\partial y} + f(x, y)F + g(x, y) = 0 \qquad (21.3)$$

containing only a single second derivative. Simple examples of this equation are relationships (14.1)–(14.3) for unsteady one-dimensional flow or for the unsteady one-dimensional heat conduction.

$$\frac{\partial T}{\partial t} = a\frac{\partial^2 T}{\partial x^2} \qquad (21.4)$$

In the age before the wide spread use of computers it was possible to solve relatively simple tasks in transport phenomena only with the engagement of outstanding mathematical skills. Only around 1960 are computers with a capacity of 16 kB RAM have become a standard gear of university laboratories. Interesting products of the mathematical treatment were the asymptotic solutions (for the limits of time and space) that served as the base for the rational definition of quantities usable for the interpretation of the results obtained. Although it is interesting to have detailed information about the distribution of temperatures in space and time, for engineering purposes it is more useful to know the heat balance and the average or the extreme values of selected variables. Today, this information could be relatively easily obtained by the methods of **numerical integration**.

21.1 Numerical integration

There is no unambiguous strategic algorithm for the mathematical solution of partial differential equations. Even the commercially available software always focuses only on a certain limited class of equations of mathematical physics, for example, the CFD (computer fluid dynamics).

The principle of numerical method of finite differences is shown using the **parabolic equation** as an example.

$$\frac{\partial F}{\partial t} = \frac{\partial^2 F}{\partial x^2} \qquad (21.5)$$

The essence of the simplest numerical solution of the equation is the resolution of x, t field into a network with an interval Δt and Δx, because it applies

$$\frac{\partial F}{\partial t} \equiv \lim_{\Delta t \to 0} \frac{F(x, t + \Delta t) - F(x, t)}{\Delta t} \tag{21.6}$$

$$\frac{\partial F}{\partial x} \equiv \lim_{\Delta x \to 0} \frac{F(x + \Delta x, t) - F(x, t)}{\Delta x} \tag{21.7}$$

and

$$\frac{\partial^2 F}{\partial x^2} \equiv \lim_{\Delta t \to 0} \frac{F(x + \Delta x, t) + F(x - \Delta x, t) - 2F(x, t)}{(\Delta t)^2} \tag{21.8}$$

It is possible from the known initial values of $F(x,t)$, $F(x-\Delta x, t)$, and $F(x+\Delta x, t)$ to calculate the value $F(x, t+\Delta t)$ in the next time interval as

$$F(x, t + \Delta t) \approx F(x, t) + \frac{\Delta t}{(\Delta x)^2}[T(x + \Delta x, t) + T(x - \Delta x, t) - 2T(x, t)] \tag{21.9}$$

In order to minimize the error it is necessary to choose the interval Δx sufficiently small. In order not to accentuate the error of approximation (21.8) of the second derivative it is necessary to shorten also the time interval Δt to secure the condition

$$\frac{\Delta t}{(\Delta x)^2} \ll 1 \tag{21.10}$$

For example, for the applicable Fourrier-Kirchhoff equation (13.11), this means a requirement

$$\frac{k\,\Delta t}{\rho\, c_P (\Delta x)^2} \ll 1$$

Although there are methods to theoretically estimate the error of a numerical solution, it is relatively easy, once the program is tweaked, to compare how the results of the calculation differ when using coarser or finer network, whether the results converge, and to which values.

The advantage of the numerical methods lies in that it is possible to choose relatively freely both the initial and boundary conditions. The disadvantage of the method of the finite differences is, however, that one network must be chosen right at the beginning of the calculation so that there is no freedom how to respect the fact that in some section of the studied field the changes in F are very significant (for example, in the boundary layer) while elsewhere the changes are only gradual. Under these circumstances, the network might be in some places insufficiently refined while elsewhere there may be unnecessarily too many knots. Modern computing is so effective that it does not constitute a problem if, after several hours of programming, several seconds or minutes is lost by unnecessary computing. If, however, problems have to be solved the simple programming of which leads to several hours of computing (this would be the case of partial differential equations of several variables) then it would be worthwhile to rationalize the solution by using a better knowledge of the mathematical nature of the problem. One such method is the method of finite elements.

21.1.1 Numerical solution using the method of finite elements

In this case, the network created by the equidistant division of the chosen coordinate system is replaced by a net of simple elements, most often linked-up triangles, the apexes of which are freely positioned in the space of the variables. An example is the network shown in Fig. 21.1. The structure is denser where a more complex distribution of the sought function is expected. Unlike the method of finite differences, this method does not study the differential balances but the integral balances of individual elements. Here too the result is a matrix, the iterative solution of which leads to a good estimate of the studied function.

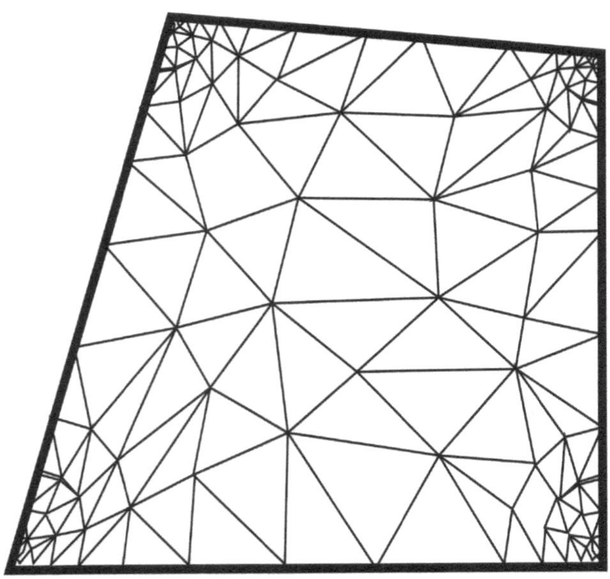

Fig. 21.1 An example of a triangular net.

While the progression of the method of finite differences can be easily programmed, the method of finite elements is substantially more complex and therefore no details will be given in this text. Most often it is offered as a commercial software (US $5000 and more for a 1-year license) that already have an expert system built-in which makes use of the existing mathematical knowledge and past experience for the optimization of the shape and the density of the computational net.

21.1.2 Monte Carlo method

This method simulates the statistical behavior of particles during the transport. To understand the concept it is necessary to imagine that the particle gets on the way carrying all the properties that it gained at the boundary of the explored region (e.g., temperature). After a very short stretch, it reaches a crossroad where it will be decided by a statistical method (using a generator of random numbers, by a draw, by the roll of a dice, or similar) which way it will proceed further. In a moving medium the particle is also swept away. This step is repeated several times until the particle does not reach a new boundary where its given property may or may not change. During this progression, the particle leaves behind a trace of its property (temperature) in each point through which it passed.

If from each point of the net forming boundary a statistically large number of particles is released then their traces can be averaged in any arbitrary point of the net, giving thus a good estimate of the temperature distribution in the field. From a programming point of view this procedure is not exceptionally demanding. It converges pretty fast but it needs a fairly long computing time if high accuracy is required.

21.2 Analytical solution

This section will remind the reader of some of the principles of solving selected partial differential equations that are useful for few simple tasks covered in this text. Primarily, these are sets with two independent variables for which it is possible to use analytical procedures at least partially.

21.2.1 The method of combining variables

In some very simple cases, it is possible to use a suitable transformation to combine the variables in such a way that the function F depends only on this only variable

$$\xi \equiv f(x)\,g(t) \tag{21.11}$$

This might happen only when the boundary conditions meet such a requirement, that is, if the boundary condition with values for $F=0$ or $F=1$ are located on the coordinates $t=0$ and $x=0$ or $x=\infty$ and the limit for $t=\infty$ is also met.

21.2.2 Eigenfunctions

The parabolic equations often have a more complex structure, for example,

$$f(x)\frac{\partial F}{\partial y} = \frac{\partial^2 F}{\partial x^2} \tag{21.12}$$

In such a case, it is customary to attempt seeking the solution again in the form of a product of two functions

$$F(x, y) = X(x)\,Y(y) \tag{21.13}$$

After substituting (21.13) in the partial differential equation (21.12) one obtains

$$f(x)\,X(x)\,\frac{\mathrm{d}Y}{\mathrm{d}y} = Y(y)\,\frac{\mathrm{d}^2 X}{\mathrm{d}x^2} \tag{21.14}$$

and by the separation of the independent variables it follows that

$$\frac{1}{Y}\frac{\mathrm{d}Y}{\mathrm{d}y} = \frac{1}{f(x)\,X}\frac{\mathrm{d}^2 X}{\mathrm{d}x^2} = C \tag{21.15}$$

Because the left-hand side does not depend on x and the right-hand side on y both sides must be equal to some constant C. Consequently, the problem reduces to the solution of an ordinary differential equation by the separation of variables

$$\frac{\mathrm{d}Y}{Y} = C\,\mathrm{d}y \tag{21.16}$$

the general solution of which (even before using the initial condition for $t=0$) comes out as

$$Y = K\exp\left(-b^2\,y\right) \tag{21.17}$$

It follows from the nature of the problem that the function F should drop with increasing y so that the constant C will be substituted by a negative number $(-b^2)$. Therefore, the equation to be solved will be an ordinary differential equation

$$\frac{d^2X}{dx^2} = -b^2 f(x)\,X \tag{21.18}$$

belonging in the class of the **Sturm-Liouville equations**.

It is quintessential that for a given assignment of the boundary conditions only a certain set of values b^2 may be appropriate, thus $\{b_1^2 < b_2^2 < b_3^2 \ldots\}$ are known as **eigenvalues**. Each eigenvalue is associated with an **eigenfunction** $X_i(x)$. The complete solution of Eq. (21.12) is

$$F(x, y) = \sum_i K_i\, X_i(x)\, \exp\left(-b_i^2\, y\right) \tag{21.19}$$

The remaining task is the determination of the value of the constant K from the initial condition. The functions X_i represent an orthonormal series and then should be valid

$$\int_{x_1}^{x_2} X_i(x)\, X_j(x) f(x)\, dx = \begin{cases} 1 & \text{for } j = i \\ 0 & \text{for } j \neq i \end{cases} \tag{21.20}$$

If the "initial" condition $F(x, y_0)$ is known then it will follow that

$$K_i = \frac{\int_{x_1}^{x_2} F(x, y_0)\, X_i(x) f(x)\, dx}{\int_{x_1}^{x_2} X_i^2(x) f(x)\, dx} \tag{21.21}$$

A particularly simple solution exists for $f(x) = 1$ when the functions $X_i(x)$ are linear combinations of the functions sine and cosine (applicable to problem dealt with in Section 14.4). Other simple cases include those resulting in the Bessel functions that are tabulated and that are included in the Microsoft Excel calculator.

In general, the solution might exist as special function that has to be back calculated. It is, however, a solution of an ordinary differential equation with boundary conditions dealt with in Chapter 20. Using the existence of the local minima b_i^2 of the functional defined by

$$b_i^2 = \frac{\displaystyle\int_{x_1}^{x_2} \left(\frac{dX_i}{dx}\right)^2 \mathrm{d}x}{\displaystyle\int_{x_1}^{x_2} X_i^2(x) f(x)\, \mathrm{d}x} \tag{21.22}$$

it is possible to arrive at an approximation of the eigenfunctions by the variation calculus.

Assumption (21.13) can be sometimes used also for elliptical two-dimensional equations but the relevant functions can be only estimated from the boundary conditions.

Appendix

Biographical notes

The following are brief references to some of the scientists whose names are linked with important quantities, relations, and concepts in transport phenomena. Although their portraits are known mainly from their ripe age some of their fundamental experiments, calculations, and inventions are dated to the time when they were 20–30 years old. Only the key summaries were published somewhat later. This fact should be an inspiration for the students.

Sir Isaac Newton (1643–1728)

English physicist, mathematician, astronomer, alchemist, and theologian. Transport phenomena owe him gratitude for the introduction of the concept of viscosity and also for a number of general quantities such as force, acceleration, momentum, and work. Of fundamental importance are Newton's three laws of classical mechanics (Fig. A.1).

Newton also developed, at the age of 23 years, independently of his contemporary Leibniz, the basic principles of the differential and integral calculus that he needed for the description of physical processes.

His main contribution is the systematic classification of all knowledge known so far, including the work of Galileo and Copernicus, which he

Fig. A.1 Sir Isaac Newton.

specified and summarized in the classical key work "Philosophia Naturalis Principia Mathematica (1687)."

Jean Baptiste Joseph Fourier (1768–1830)

French mathematician and physicist

In 1822, Fourier summarized his work in a treatise entitled "Theorie analytique de la chaleur". He formulated and, in some situations, solved the differential equations of convective heat transport for which he developed some original mathematical methods. The well-known ones of which are Fourier transformation, Fourier series, and others. He also discussed the existence of radiation heating. Fourier is also given credited with the phenomenon "glass house effect" (Fig. A.2).

Fig. A.2 Jean Baptiste Joseph Fourier.

Daniel Bernoulli (1700–82)

Swiss physicist and mathematician

In his crucial work Hydrodynamica (1738), he summarized the results of all of his experimental and theoretical studies, including the Bernoulli equation which describes the flow of fluids. (Some ideas from the domain of hydrodynamics were simultaneously worked on by his father Johann with whom Daniel argued about the priority.) In the appendix to his book, Daniel created the first kinetic theory of gases which he considered to be "elastic liquids" made of particles. To describe their behavior he used the theory of probability. From the deliberations about the microscopic structure of gases he developed macroscopic equations describing their behavior during temperature and pressure changes (Fig. A.3).

Fig. A.3 Daniel Bernoulli.

Leonhard Euler (1707–83)

Swiss mathematician and physicist, disciple of Bernouli, most of his life was spent in Saint Petersburg.

Euler is known mainly for his achievements in the field of differential calculus and the theory of graphs. He has also developed a large number of new concepts and symbols, particularly in mathematical analysis. He is mainly known for his work in mathematics although his contributions to mechanics, particularly to hydrodynamics, and also to optics and astronomy, are significant (Fig. A.4).

Fig. A.4 Leonhard Euler.

Claude-Louis Navier (1785–1836)

French physicist and engineer

In 1821, he mathematically formulated the theory of elasticity and he defined the modulus of elasticity. Two years later, he derived the equation of motion of a viscous fluid. As an engineer, he is renowned for the design and construction of bridges (Fig. A.5).

Fig. A.5 Claude-Louis Navier.

Augustin Louis Cauchy (1789–1857)

French mathematician, physicist, and engineer

His work in mathematics is very extensive. It encompasses many tools of mathematical analysis that are in use. Well known is his solution of differential equation having only an initial condition. Another task treated in this book is his differential equation of momentum balance in terms of stresses (Fig. A.6).

Fig. A.6 Augustin Louis Cauchy.

Jean Léonard Marie Poiseuille (1797–1869)

French physician and physiologist

While studying the bloodstream he derived (1838) the relationship for the flow of blood in capillaries (valid generally for laminar flow in tubes of circular cross section) (Fig. A.7).

Fig. A.7 Jean Léonard Marie Poiseuille.

Sir George Gabriel Stokes (1819–1903)

British physicist and mathematician

Independently of Navier he formulated the "Navier-Stokes" equations of fluid flow and the "Stokes law" for the motion of a sphere in a viscous medium. In mathematics, he contributed to the theory of series and he formulated "Stokes theorem" from vector calculus. He also dealt with physical optics (Fig. A.8).

Fig. A.8 Sir George Gabriel Stokes.

Gustav Robert Kirchhoff (1824–87)

German physicist, chemist, and mathematician

Kirchhoff's laws (1845) are universally valid balances used in electrical engineering as well as for the description of branching of heat in parallel or serial networks. They are, however, applicable to flow and diffusion. Kirchhoff is best known for his work in spectral analysis. He also explained the relationship between emission and absorption of radiation (1859). He also defined the concept of the "black body" (Fig. A.9).

Fig. A.9 Gustav Robert Kirchhoff.

Franz Grashof (1826–93)

German engineer

He developed methods for the description of the flow of steam. He also significantly contributed to the understanding of free convection (Fig. A.10).

Fig. A.10 Franz Grashof.

Adolf Eugen Fick (1829–1901)

German physician and physiologist

He described the laws governing diffusion (Fick's law) and he defined diffusivity (1855). In medicine, he developed the principle used to quantify cardiac performance (1870) (Fig. A.11).

Fig. A.11 Adolf Eugen Fick.

Osborne Reynolds (1842–1912)

English physicist and engineer

He is the author of the famous experiment demonstrating the transition from laminar to turbulent flow. He also dealt with the mechanics of gases and of granular materials. As an engineer, he is renowned for the design of ship propulsion (Fig. A.12).

Fig. A.12 Osborne Reynolds.

Ludwig Boltzmann (1844–1906)

Austrian physicist, the founder of statistical physics

His life work was thermodynamics. Along with James Maxwell he founded statistical mechanics. He presented entropy as a macroscopic quantity. He also supported the atomistic interpretation of mass.

Along with his teacher Jozef Stefan he derived the law governing the transfer of heat by radiation. He showed that the intensity of the radiation depends on the fourth power of the thermodynamic temperature of the emitter (Fig. A.13).

Fig. A.13 Ludwig Boltzmann.

Nikolay Yegorovich Zhukovsky (1847–1921)

Russian mathematician and physicist

He is the founder of modern hydrodynamics and aeromechanics. He discovered the physical law governing buoyancy acting on the wings of an aircraft. He developed mathematical theories for all components of an aircraft (Fig. A.14).

Fig. A.14 Nikolay Yegorovich Zhukovsky.

Ludwig Prandtl (1875–1953)

German physicist and engineer

He carried out a key innovation in solving the Navier-Stokes equation when he formulated in 1904 the concept of "the flow with a small influence of friction." This concept was based on the empirical observation that the influence of friction on the resistance of bodies bypassed by the flowing media is limited to a thin layer that he called "the boundary layer." For its description he introduced the "Prandtl equations" (Fig. A.15).

Fig. A.15 Ludwig Prandtl.

Theodore von Kármán (1881–1963)

Hungarian physicist first active in Germany and from the year 1929 in USA

He further mathematically developed the concept of boundary layer and applied it for the development of aircraft wings and propellers. Later, he also contributed to the development of jet engines (Fig. A.16).

Fig. A.16 Theodore von Kármán.

Ernst Kraft Wilhelm Nusselt (1882–1957)

German engineer

In 1910, he significantly contributed toward the solution of natural heat convection in laminar flow in tubes (Graetz solved this task for the plug flow already in 1883). He also solved the problem of film condensation on a vertical wall (Fig. A.17).

Fig. A.17 Ernst Kraft Wilhelm Nusselt.

Andrey Nikolajevich Kolmogorov (1903–87)

Russian mathematician

Founder of the modern theory of probability and the complexity theory of algorithms. He also worked in the area of topology, logic, Fourier series, turbulence, and classical mechanics (Fig. A.18).

In 1941, he published his fundamental work which provides the mathematical tool for the interpretation of turbulent flow based on the hypothesis of local isotropy.

Fig. A.18 Andrey Nikolajevich Kolmogorov.

Alexey Vasilievich Luikov (1910–74)

Russian mathematician and physicist

He contributed significantly to the research of heat and mass transfer beginning with the imbalance thermodynamics and ending with the numerical methods of solving nonlinear equations (Fig. A.19).

Fig. A.19 Alexey Vasilievich Luikov.

Index

Note: Page numbers followed by *f* indicate figures and *t* indicate tables.

M

N

O

P

Q

T

U

V

W

Y

MIX
Papier | Fördert
gute Waldnutzung
FSC® C083411

Kolibri 360 GmbH
Ferdinand-Jühlke-Straße 7
99095 Erfurt, Deutschland
produktsicherheit@kolibri360.de